Environmental Science and Engineering

Environmental Science

Series editors

Rod Allan, Burlington, Canada
Ulrich Förstner, Hamburg, Germany
Wim Salomons, Haren, The Netherlands

For further volumes:
http://www.springer.com/series/3234

Vishwambhar Prasad Sati

Towards Sustainable Livelihoods and Ecosystems in Mountain Regions

Springer

Vishwambhar Prasad Sati
Geography and Resource Management
School of Earth Sciences
Mizoram University
Aizawl, Mizoram
India

ISSN 1431-6250
ISBN 978-3-319-03532-1 ISBN 978-3-319-03533-8 (eBook)
DOI 10.1007/978-3-319-03533-8
Springer Cham Heidelberg New York Dordrecht London

Library of Congress Control Number: 2013955052

Printed on acid-free paper

Springer is part of Springer Science+Business Media (www.springer.com)

Preface

Diversifying and enhancing livelihood options for reducing poverty and securing livelihoods is an essential activity. It has greater relevance in areas where subsistence agricultural practices dominate the economy. About 70 % of the world's economy is based upon the cultivation of subsistence cereal crops. This practice is a major characteristic feature of mountain regions. Mountains constitute about 20 % of the total earth's land-surface and comprise of underdeveloped economy. Here, agriculture and its allied activities are the main occupation of the people. Agriculture is largely carried out on narrow patches of terraced fields. Cultivation of subsistence cereals dominates the cropping pattern. Production and per ha yield of cereal crops is considerably low and does not meet even the two-times food requirements of the poor and marginal farmers. The case study area, 'the Alaknanda Basin' constitutes one of the major sub-systems of the Ganges. It has five vertical landscapes (zones) from the valleys to the highly elevated Himalayan ranges, i.e. the river valleys, mid-altitudes, highlands, alpine meadows and the perpetual snow-clad mountain peaks of the Himalayas. Human settlements and high population density are largely concentrated in the mid-altitudes and in the valleys, while the highlands are sparsely populated. The highland areas are used for grazing animals through seasonal migration mainly during the summer. The alpine meadows remain covered with snow for up to 6 months during the winter, and during the summer an abundance of flowers and medicinal plants grow naturally. Medicinal plants are collected by the local people and widely used for health care through the traditional system.

The Himalayan region is ecologically fragile, tectonically and seismically active, geologically unstable, economically backward and geographically remote. So far, the waves of modern civilization could not influence the whole region because of its remoteness and distance from mainstream development. High growth in population and less availability of arable land further accelerates the human pressure on the agricultural land. Intensive agricultural practices are carried out on the fragile and narrow terraced fields, which are located in the mid-altitudes and the highlands. In the valley regions, wherever narrow patches of terraced fields are found, agricultural practices are carried out. Extension of arable land towards the fragile mountain niches poses severe threats to the landscape stability as the entire region is highly vulnerable. Suitability in the agro-climatic conditions for growing cash crops such as fruits and off-season vegetables in particular, leads to sustainable development of the Himalayan region. Similarly, an abundance of economically viable forests with rich agro-biodiversity manifests to diversifying livelihoods at a larger scale. Multiple uses of timber and non-timber forest products such as bees and bee keeping, natural dyes, bamboo and bamboo-based products, herbs, wild fruits and their products, oak bark and its products, and a number of other products can enhance the livelihood options of the people.

The book constitutes ten chapters. In the Chap. 1, research problems, questions, objectives and hypotheses are illustrated. Literature review, methodology and tools and techniques of the research, which were used to carry out this study, are well elaborated in this introductory part. Chapter 2 deals with the geographical background. It further elaborates the climatic conditions and natural resources availability. Chapter 3 discusses *Socio-Economy and Population Profile* of the study area. *Sustainable Livelihoods: Diversifications and Enhancements* are discussed in Chap. 4. Agricultural intensification and diversification through cultivating cereals and cash crops and use of timber, non-timber and biodiversity resources are widely illustrated in this chapter. *Livelihood Analysis* viz. income–expenditure analysis of the households is carried out in Chap. 5. Chapter 6 examines the potential of development of tourism and hydroelectricity in the Himalayan region and illustrates how they are useful in terms of enhancing livelihoods. Case studies of the Khanda Gad sub-watershed of the lower Alaknanda basin and the Kewer Gadhera sub-watershed of the Pindar basin are discussed in Chap. 7. Household level survey of the villages of these two sub-watersheds was carried out. Chapter 8 discusses *Mountain Ecosystems* and impacts of anthropogenic activities on it. *Sustainable Mountain Development and Livelihood Enhancement: Challenges and Opportunities* are examined in Chap. 9. Finally, conclusions and suggestions are given in Chap. 10 that also describes the major problems and prospects of sustainable livelihoods. Illustrations exhibiting the natural and cultural landscapes of the Himalayan region and appendices that give numerical information further enrich the manuscript.

This work is a result of the ICSSR Fellowship, which was awarded to me as 'ICSSR Fellow' during 2008–2009. This fellowship was carried out in the Department of Geography at the HNB Garhwal University Srinagar, Uttarakhand. The Mizoram University, Aizawl provided publication grant to this work. I acknowledge my gratitude to the ICSSR, New Delhi for awarding this fellowship. I am thankful to the Department of Higher Education, the Government of Madhya Pradesh for sanctioning study leave. I am indebted to the Department of Geography, HNB Garhwal University for allowing and providing me all the necessary facilities to commence this study. I am also grateful to the Mizoram University for providing publication grant to publish this book. The academicians and the scholars working in mountain regions are greatly thankful as their work is well cited in this book. I acknowledge the contribution of Mrs. Lekha Bhat for taking the initiative to edit this book. I am indebted to my family members; Dr. Nirmala Sati and Vishwani Sati for the valuable support that they have extended from time to time. I dedicate this book to my grandmother late Smt. Savitri Devi, grandfather late Shri Govind Ram Sati, mother late Smt. Saradi Devi and father late Shri Shiv Dutt Sati.

Aizawl, 2013 Vishwambhar Prasad Sati

About the Book

Sustainable livelihoods and Ecosystems are comprehensive and burning issues in the wake of high growth of population, low production and per ha yield of crops, and depletion of biodiversity resources. Mountainous regions of the world are facing the menace of poverty, food insecurity and malnutrition. Further, tremendous growth in population and slow pace of development together push most of the population to live below the poverty line. Traditionally depending upon cultivating subsistence crops for food requirement, people living in the mountainous regions cannot produce sufficient food grains to manage their livelihood smoothly. The Himalayas, one of the biodiversity hotspots of the world, has an abundance of natural resources—land, water, and forest, the life sustaining layers. The geo-environmental conditions—climate and landscape further enhance the possibility of sustainable livelihoods through eco-tourism, harnessing water resources, utilising forest and its products sustainably. Diversifying agricultural practices through cultivating cash crops and enhancing livelihood options through extensive use of timber and non-timber-based forest products thus, can assist to eradicate poverty and ensure food security. There are ten chapters comprising Introduction, Geo-Environmental Setting, Socio-Economy and Population Profile, Sustainable Livelihoods: Diversification and Enhancement, Livelihood Analysis, Development of Tourism and Hydro-Electricity, Case Study, Mountain Ecosystems, Sustainable Mountain Development and Conclusion. This book will be highly significant to all stakeholders working for the sustainable livelihoods in mountain regions.

Contents

Abbreviations

AEZ	Agri Export Zones
APEDA	Agriculture and Processed Food Products Development Authority
ADB	Asian Development Bank
ARB	Alaknanda River Basin
BDF	Bhararisen Dairy Farm
CDPCUL	Chamoli District Dairy Production Cooperative Union Limited
COI	Census of India
DID	Department of International Development
FA	Forest Act
FAO	Food and Agricultural Organization
FDA	Forest Development Agency
GBPIHED	Govind Ballabh Pant Institute of Himalayan Environment and Development
H and MP	Herbs and Medicinal Plants
HAPPRC	High Altitude Plant Physiology Research Center
HDR	Human Development Report
HDRI	Herbal Research and Development Institute
HMS	Himalayan Mountain System
HNBGU	Hemwati Nandan Bahuguna Garhwal University
HYV	High Yield Variety
IC	Indian Currency
ICAR	Indian Council of Agricultural Research
ICHR	Indian Central Himalayan Region
ICIMOD	International Center of Integrated Mountain and Development
ICSSR	Indian Council of Social Science Research
IHR	Indian Himalayan Region
IIEs	Integrated Industrial Estates
IPCC	Inter-Governmental Panel of Climate Change
IDD	Iodine Deficiency Disorder
IMDP	Intensive Mini Dairy Project
JMS	Journal of Mountain Science

KGSW	Kewer Gadhera Sub-Watershed
KGSW	Khanda Gad Sub-Watershed
LSI	Livelihood Sustainability Index
LODI	London-based Oversees Development Institute
MPCA	Medicinal Plant Conservation Area
MRD	Mountain Research and Development
NAGI	National Association of Geographers, India
NAP	National Afforestation Programme
NAPCC	National Action Plan on Climate Change
NGOs	Non-Governmental Organisations
NMSHE	National Mission for Sustaining Himalayan Ecosystems
NORAD	Norway Research and Development
NSP	Nauti Sub-Projects
NTFP	Non-Timber Forest Products
PIB	Public Investment Board
PRA	Participatory Rural Appraisal
RLD	Rural Livelihood Department
SC	Scheduled Caste
SHN	Shubhas Herbal Nursary
SL	Sustainable Livelihood
ST	Scheduled Tribes
TF	Terraced Farmland
ULDB	Uttarakhand Livestock Development Board
UNCBD	United Nations Convention of Biological Diversity
UNCCD	United Nations Convention to Combat Desertification
UNCED	United Nations Conference on Environment and Development
UNDP	United Nations Development Programmes
UNEP	United Nations Environment Programmes
UNFCCC	United Nations Framework Convention on Climate Change
UNU	United Nations University
UTDB	Uttarakhand Tea Development Board
VFDC	Village Forest Development Committee
VFDC	Valley Flower Development Corporation
WCED	World Commission on Environment and Development
WLA	Wildlife Act

Chapter 1
Introduction

The term 'livelihood' refers to enough food and cash to meet basic needs (Chambers 1988) and poverty reduction also mentions those 'livelihoods' under the basic definition of poverty. Chambers and Conway (1991) explained, 'a livelihood includes the capabilities, assets (financial, physical, human, natural resource, and social) and activities required for a means of living'. Here assets include both material and social resources, or both tangible and intangible resources. Poverty indeed is the result of a lack of these assets. The poor may not have access to these assets and this could be because of physical constraints like remoteness or resource scarcity (Asian Development Bank). Scholars like Jeganathan have observed that livelihood aims to generate adequate resources, both cash and non-cash, which is a set of economic activities by utilising human and material resources. Livelihood includes the access to resources, materials and services into the content (Ellis 2000). Besides material and economic supplies, livelihood is also about management of social relationships, personal and group identity and the interrelation of these tasks to each other. Livelihood can also include creativities, new opportunities, people's attempts, willingness and capabilities to cope with shocks, risks and stresses, such as natural disasters, epidemics such as HIV/AIDS, financial crisis or conflicts and competition both at national and international levels (UNDP 2006). These intensive definitions of livelihood also reflect the multi-goals of poverty reduction, in which the development of human and social capital increases the possibility of poverty reduction. In this way, livelihood is a very comprehensive term that deals with all issues related to human needs and capacities.

Livelihood is sustainable when it can cope with and recover from stresses and shock; can maintain or build on available capabilities and assets; and does not undermine the natural resource base. Sustainable livelihood (SL) is the long-term goal for poverty reduction that gives adequate importance to development from different levels, scale and sectors. It is regarded as a conceptual framework, can be used to assess and analyse sustainability, capability, security and resilience of livelihoods in different scales under different contexts (Scoones 1998a, b). SL tries to work out integrated solutions under the complicated and flexible political and economic environment.

V. P. Sati, *Towards Sustainable Livelihoods and Ecosystems in Mountain Regions*, Environmental Science and Engineering, DOI: 10.1007/978-3-319-03533-8_1,

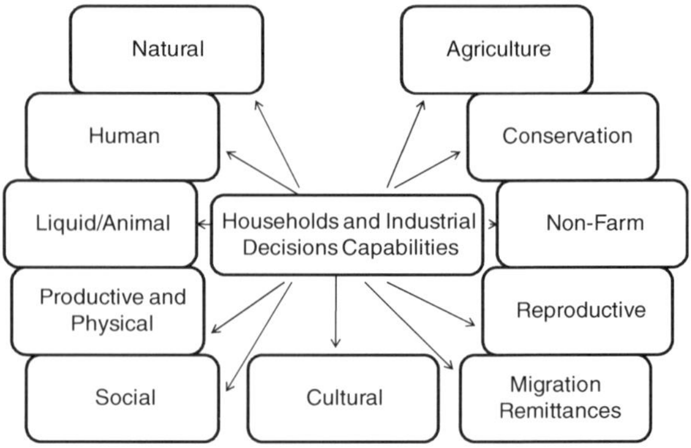

Fig. 1.1 Livelihood strategies: assets, decision and activity portfolios

Mountains occupy 24 % of the global land surface and are home to 12 % of the global population. In addition, another 14 % of the global population lives adjacent to mountain areas. Most of the people living in mountain regions are economically vulnerable and socially backward (ICIMOD 2007). Mountains are sources of water, energy, agricultural and forest products, centres of biological and cultural diversity; besides this tourism and pilgrimages are the other important aspects upon which the livelihoods of the mountain people are dependent (Ives and Messerli 1989). Further, among the mountain communities cultural and traditional diversity is also high. Mountains have ecological, aesthetic and socio-economic significance both as natural ecosystems and as people's living place. 10 % of the world population's livelihoods and well-beings rely directly on mountainous resources (UNU 2002). For sustainable development of the mountain regions, various methodologies are required, primarily to deal with development issues under its complex ecological and agricultural systems. It is the interactivity between social and ecological systems, and between human and nature (Kemp and Martens 2007). Due to the specificities and peculiarities of the mountain regions, there is a need to have specific and targeted policies. The livelihood of mountain people in many countries of the world are still lagging behind in their development processes (Figs. 1.1, 1.2, 1.3, 1.4, 1.5).

Enhancing and diversifying livelihood options in mountain regions need attention against the backdrop of framing an integral strategy for coping with physical hazards and food insecurity. It must also seek ways to improve livelihood and enhance economic growth through which increased security—physical, economic and social can be obtained. The people of mountain regions are entirely dependent on mixed agriculture systems, which include farming of subsistence cereal crops and animal husbandry. Low production and productivity (per ha yield), home consumption of produced materials and limited access to market are the characteristics of mountain systems. The potentials to avail sustainability through enhancing and diversifying livelihood options, within the context of vulnerability and fragility of mountain terrain, have largely remained unexplored by mountain habitants. The impact of this is that the phenomena of poverty, food

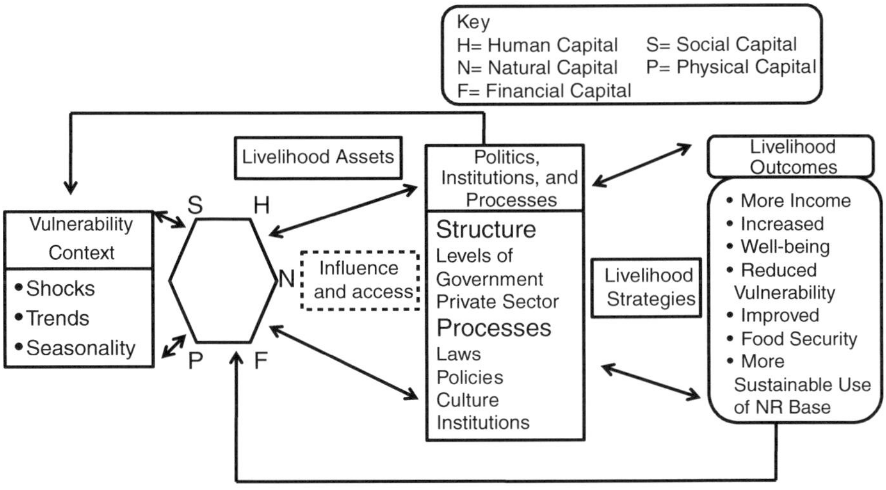

Source: DFID, 2001

Fig. 1.2 Sustainable livelihood framework

Fig. 1.3 Livelihood components and flows

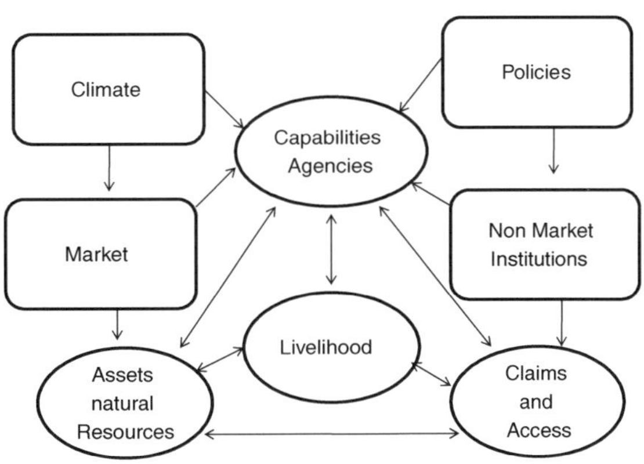

insecurity and malnutrition are still prevalent in these regions. In addition to this, high population pressure and high dependency on forest resources has led to severe environmental degradation.

Most of the mountain areas have not been able to adequately harness their unique resources to improve livelihoods because of inadequate and unfavourable policies. Harnessing mountain niches appropriately through (i) better management of natural resources; (ii) application of technologies; (iii) new methods of production and exchange to generate employment and income opportunities in the mountain regions is required. However, the cultivation of off-season vegetables, fruits, medicinal plants along with the appropriate use of non-timber-based forest

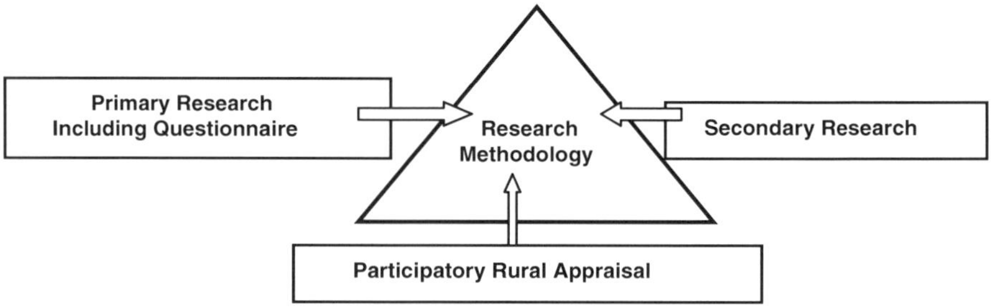

Fig. 1.4 Sustainable livelihood research design (A)

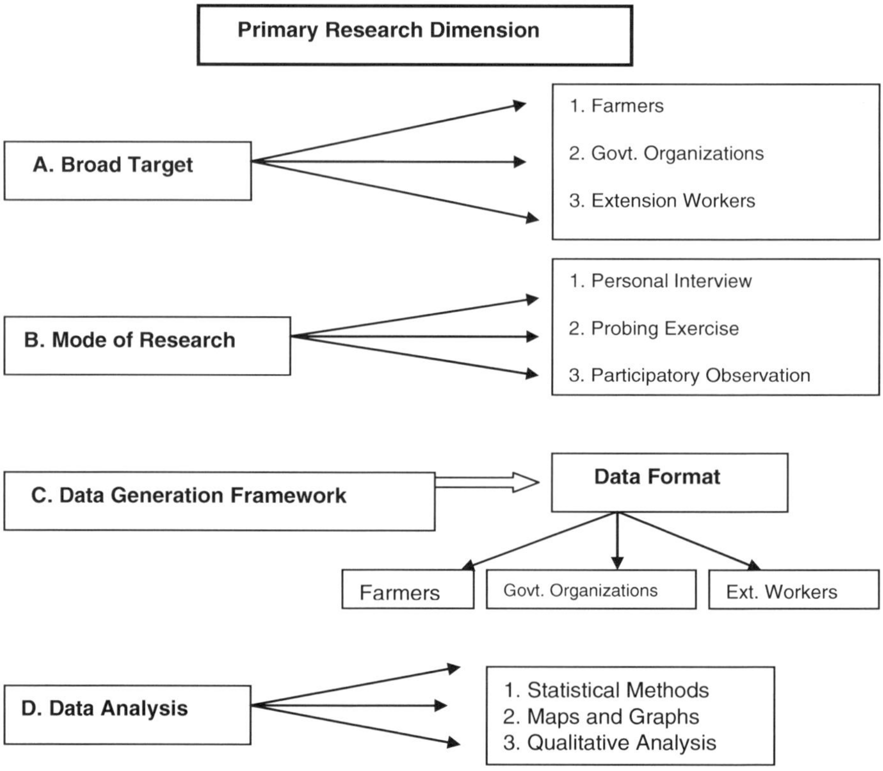

Fig. 1.5 Sustainable livelihood research design (B)

products of the mountains demonstrate their high potential to provide viable bases for households to rise above poverty and subsistence.

In mountain areas, livelihood options are often linked to a range of economic activities, products and productivity of cereal farming, natural assets of mountains as well as economic and human assets. Harnessing mountain resources for hydroelectricity, tourism and for the production of food/non-food products along with conserving resources to generate valuable environmental services can create new employment and income opportunities. Human resource development is also vital for the all-round development of mountain regions.

Agriculture and its associated activities continue to play a major role in the economy of mountain regions. It is a major source for providing employment to the rural populace. In India, for the past several centuries, agricultural practices are avenues of subsistence and agriculture is sustaining the livelihood of rural households. Over the years, unprecedented increasing trend of population pressure on land—both—for employment and livelihood has led to a situation where the land–man ratio has considerably declined. Further, increasing fragmentation of agricultural holdings and inequalities in the distribution of arable land among farming households, have caused decline in per household income. Despite increasing application of labour-saving farm production technologies, the agriculture sector is hardly in a position either to provide gainful employment opportunities at pace with the increasing workforce or to sustain the livelihood of rural households. Carrying out a large-scale diversification and modernisation of economic system through initiating industrialisation and large-scale production system is severely limited in hilly and mountainous areas because of a number of factors—such as the availability situation of limited environmentally sensitive resource bases and the spread of usable resources across different and inaccessible areas. Factors such as inaccessibility to markets/modern inputs/technology, deficient infrastructure and high transport costs are also leading to non-competitiveness of products (Mehta 1996).

Geographical and Economic Constrains to Sustainable Livelihoods in Mountain Regions

Mountain regions have been regarded as the most fragile, remote and marginal ecological systems on the earth. As they are lagging behind in infrastructure facilities, returns from investment are tremendously low. As a result, in many cases, development of mountainous areas is experiencing a minor and passive mode that is dependent on external aids or strong governmental interventions (Klaus 1983). Some scholars believe that in the agricultural society, the differences in people's living standards in mountain areas and plains were not so remarkable; but in a globalised world, mountain areas became more vulnerable and widened the regional imbalances (Klaus 1983; Guojie 2007). Besides, natural disasters such as earthquakes, flash floods and the risk of climate change are all drivers that increasingly threaten the capacity of mountainous sustainability. This in turn influences mountain communities and even lowland populations.

The Indian Central Himalayan Region (ICHR) is one of the richest regions in terms of biodiversity resources, but is amongst the poorest in terms of economic growth and development. The livelihood is largely dependent on the biomass-based production and on the use of natural resources—forest, grassland, fresh water etc. Agriculture in this region is based upon century-old practices, which are carried out mainly on narrow patches of terraced fields and characterised by the dominance of subsistence cereal farming, which is the main occupation of the populace. Rice, wheat, barley, millets, oilseeds and pulses are the main cereal crops and the economic viability of these crops is insufficient even to meet the daily basic requirements. The scope of modernisation and expansion of agriculture is negligible. Fragility of the terrain further reduces the scope of implementation of modern innovations in the field of agriculture. Mounting pressure of human population on arable land and high man–land ratio compelled farmers to leave the land abandoned. This process led to low soil fertility and land degradation. Apart from adverse climatic stress, increased human population and the insatiable demand for more natural resources including land, forest and food are the major factors contributing to natural resources depletion and losses in biodiversity (Arimoro et al. 2002; Okali 1985). Furthermore, inaccessibility and the low level of infrastructural facilities are significant constraints to generating opportunities. As a result, the resources of the region have remained grossly underutilised and undervalued. These natural and man-made disadvantages manifest in the form of poverty, malnutrition and food scarcity. As a coping strategy to these severe impediments, a large proportion of the population has migrated to the foothills of the Himalaya and the Ganges plains. There are instances from the Garhwal Himalaya where about two-thirds of the total families in a village emigrated during the period 1972–1988 (Maikhuri et al. 1995). Unsustainability of agriculture and lack of other economic opportunities together with higher literacy rate in the hill region lead to very high rate of out-migration of the youth in search of jobs (Khanka 1984).

In the Himalayan region, the impediments related to the development processes are not only due to the adverse geo-environmental conditions, but also due to the policies of the various governments (both State and Central) and development agencies. Lack of systematic planning was accentuated by inadequate and improper coordination between different governmental departments. These programmes paid scant attention to the needs and requirements of various infrastructural facilities for sectoral schemes, inefficiency of the administration in implementation of programmes, a multiplicity of programmes to meet the same goals and faulty criteria for identifying beneficiaries in assistance linked to employment generation (Mehta 1990).

The landscape, geo-environmental conditions and availability of natural resources have heterogeneity in all respects and this can be seen on the agrarian system, occupation, working potential and migration. Other aspects such as low availability of agricultural land, harsh environmental conditions, inaccessibility of forestland and instability of terrain affect the livelihood of the Himalayan people (Sati and Kumar 2004). With increasing needs as well as pressure of population, traditional farming has become unsustainable both economically and ecologically (ICIMOD 1996). Further, expansion of agriculture on marginal land and declining crop yields are considered to be major unsustainable trends in the Himalayas (Eckholm 1979; Jodha 1990).

Horticulture—cultivation of fruits and off-season vegetables plays a vital role for the environmental sustainability and economic development of the region (Sati 2004a, b, c, d). The present vegetable production of 90.8 million tonnes is to be raised to 250 million tonnes by 2024–2025 (ICAR 2002). Promotion of protected cultivation of vegetables is another potential approach [see Singh (1998) and Singh et al. (1999)]. There are different ways and means to achieve this target, which include among others, bringing additional areas under vegetable crops, using hybrid seeds and use of improved agro-techniques. Cultivable wasteland may have potential for the development of horticulture, whereas the uncultivable wastelands do not have such potential (Singh 1991). In addition, non-timber forest products such as medicinal plants/herbs, essential oils, fibres/silks, natural dyes/organic products, bamboo/bamboo products, bee/bee products and enterprise-based pollination services can provide the bases for increasing incomes and improving livelihoods. It is commonly observed that the ecological conditions of the region are more suitable for fruit cultivation rather than cereal farming (Atkinson 1889). Along with the cultivation of fruits, off-season vegetables and tea cultivation will boost the regional economy (Sati and Kumar 2004). In the Himalayan region from the valley regions to the northern border, sub-tropical humid and bio-climatic conditions change step by step into temperate, sub-temperate and alpine zones (Atkinson 1889), which are very useful for the cultivation of fruits and vegetables. Likewise, mountain tourism, hydroelectricity and other renewable energy sources may be other resources for enhancing livelihoods. These niche-based mountain products and services hold some of the keys for helping mountain people diversify and enhance their livelihood options while reducing dependency on traditional occupations and the environment.

The development process must also ensure that communities, (especially disadvantaged groups) and geographically excluded areas are not left out of mainstream development. Mountain areas are highly diverse in renewable natural resources and environmental services. The diversity is helping to reduce internal competition in mountain areas and partially offsetting the physical vulnerability of the fragile mountain environment. It requires a highly decentralised areas-based approach (Papola 1996), which has to be distinct not only from approaches for the plains, but should also differ significantly from area to area within the hill region. The biodiversity of the mountains provides important values of agriculture, medicine, food security and industry besides spiritual, cultural and aesthetic and recreational values. In order to meet the present and future challenges meeting sustainability criteria, traditional systems need to be adapted in ways that will enhance crop yields but not at environmental and social costs (Ramakrishnan et al. 1993).

Literature Review

Diversification of livelihoods is a prime concern in the wake of food security and poverty reduction. A number of different sustainable livelihood approaches have emerged, but most rely on analytical frameworks which owe significance to sustainable livelihood programmes (Scoones 1998a, b). The Rural Livelihoods Department (RLD) of the United Kingdom's Department for International Development (DID) (DFID 2000) funded four parallel studies between 2000 and 2003 to identify how the livelihoods of the rural poor can be enhanced. These were conducted in southern Africa, eastern Africa and in South Asia. The study aimed to identify innovative ways of enhancing livelihoods, and to promote exchange of experience among donors and governments across the regions. One of the four studies, conducted by the London-based Overseas Development Institute (LODI) was concerned with livelihood options.

The studies have stressed that traditionally emphasised determinants of production are not the only ones that are important. Reardon et al. (2001) have stressed that in addition to these traditionally emphasised determinants of productivity and capital investments, importance should be given to non-farm income, soil conservation investments and market infrastructure improvements. Similarly, Guyer and Lambin (1993) have noted that agricultural practice is developing in a dynamic fashion ahead of population pressure, largely due to market responses. Schelhas (1996) and Roumasset et al. (1979) have suggested that risk and the perception of risk are key factors influencing the nature and timing of intensification and diversification decisions. A recent study on the evolution of commercial vegetable gardening from 1971 to 1988 in the Philippines (Eder 1991) found that there had been a two–threefold increase in the labour invested per unit of land. Binswanger and Khandker (1993) have taken this aspect further in their examination of how financial institutions and interest rates determine investment, input and output decisions in India. The importance of understanding the broader political and economic issues to explain the presence (or absence) of sustainable livelihoods has been highlighted by a number of studies. Conelly (1994) sought to explain why farmers of Rusinga Island in Kenya have abandoned intensive agricultural practices. Similarly, in Usagara and Tanzania, a change from cotton to rice and the increased use of manure as fertiliser, can be explained in part by the changing factor prices, related to the removal of subsidies on fertilisers (ICRA 1990). Kelly et al. have observed that in Senegal, the use of fertiliser has declined over the years due to structural adjustment. A study by Diagana and Kelly (1996) examines how the profitability of main crops has changed and affected the choice of crop mix and technology; they argue that devaluation has not encouraged farmers to sustainable patterns of intensification characterised by high fertiliser use.

The complex of ideas, principles and analytical tools, which has been labelled the sustainable livelihoods approach, has acquired extraordinary prominence in developing thinking and activity within a very short time. Over the last few decades, our understanding of livelihoods and poverty has considerably changed/

modified and this has its implications in the way in which we define research and analyse these concepts. Walker and Ryan (1990) considered agricultural household economics in villages to enhance 'the understanding of the dynamics of agricultural development in one of the poorest rural regions of Asia'. In sub-Saharan Africa, this diversification is frequently interpreted as a response to the difficulties that poor households face in the context of structural adjustment and liberalisation (Bryceson 1999; Ellis 1998 and 2000). The ways in which households in India are responding to the processes of liberalisation are the subject of intense debates and continue to be disputed (Meenakshi and Ray 2002; Datt and Ravallion 2002). Research elsewhere has shown that diversification is neither necessarily a strategy pursued by poor people, nor is it just about coping. For some people it can help in mitigating risk or coping with vulnerability where risk remains high and in setting poor people on a cumulative path towards greater livelihood success (Davies 1996). The above-cited studies explain the intensification of agriculture for sustainable livelihoods and focus on the current trend of diversification of agricultural crops. Likewise, many other studies are currently emphasising on agricultural intensification, diversification and livelihood enhancement worldwide.

The Himalayan region is ecologically fragile, tectonically and seismically active, geologically sensitive, geographically remote and economically underdeveloped. Similarly, in terms of theoretical and practical research work, on various aspects, it has lagged behind. Although some noteworthy works have been done in the Himalayan region on farming systems, development aspects and natural resources management, thorough research is yet to be initiated. Many scholars of the two universities located in the region (Garhwal and Kumaon), and some research institutes such as 'Govind Ballabh Pant Institute of Himalayan Environment and Development' (GBPIHED) and 'High Altitude Plant Physiology Research Centre' (HAPPRC) presented noteworthy contributions through various research projects and Ph.D. works. 'Uttaranchal: Dilemma of Plenties and Scarcities' (2004) written by the present author, is a noteworthy contribution in the field of evaluation of natural resources, their utilisation pattern and strategic planning for speedy development of the Uttarakhand state. A book on "Horticultural Development in Hills: A Case for the Alaknanda Basin" (Ph. D. thesis of the present author) is also a notable contribution. 'Natural Resources Management and Sustainable Development of the Pindar Basin, Himalaya', by the present author (2008) and 'Pilgrimage Tourism in the Garhwal Himalaya', by Sati and Mansoori (2009) are some other worth-mentioning contributions in the field of livelihood enhancement in the Himalayan region.

Research papers on 'Problem of Agricultural Ecology and its Management in the Pindar Basin', 'Cropping Pattern in the Hill Environment of Garhwal Himalaya', 'Pattern of Resource Utilization and Development in Hills: A Case for the Pindar Basin of Garhwal Himalayan', Fruit Cultivation in Hills: A Case for the Alaknanda Basin', 'Vertical Zonation of Horticultural Farming in the Alaknanda Basin', 'Systems of Agricultural Farming in the Uttaranchal Himalaya', 'Natural Resource Conditions and Economic Development in the Uttaranchal Himalaya, 'Intra-Montane Mobility', 'Agricultural Diversification', 'Forest Biomass Flow

Systems', 'Traditional Farming and Sustainability Issue', 'Natural Resources Management and Food Security', 'Cultivation of Medicinal Plants', 'Natural Hazards in the Himalaya', 'Migration and Agrarian Changes', 'Diversifying and Enhancing Livelihoods', 'Sustainable Livestock Farming' and many other papers were published in peer-reviewed international journals and several leading research journals of India by the present author.

On the basis of the above literature, it becomes evident that despite the increasing volume of research works in the Himalayan region, no specific work was carried out on sustainable livelihood, food security and resource management. This study presents a comprehensive picture on agriculture diversification and options for livelihood enhancement and suggests detailed policies and planning. The aspects related to their implementation in a rational manner are also discussed in detail.

Conceptual Framework

The hill farming system is complex where crop production, animal husbandry and forestry are intricately linked. Together, they determine the living standards of farm families, income and employment levels and affect their surrounding environment. Forest lands provide fuel wood, fodder and timber. Croplands provide food, fodder and crop residue. However, croplands also require manure, which is available from cow dung and litter from the forestlands. Thus, livestock connects these land resources by converting fodder into drought power and dung nutrient, in addition to providing food and income to households.

Large parts of the Himalayas are still isolated pockets and are remote from the mainland; as a result they are insulated from market forces. Transport network and communication facilities have not reached many parts of the region. Movement of people and farm products to urban centres is difficult. Tourism can play an important role in some of the natural picturesque areas of the basin. Development of educational facilities is a major aspect, which will surely lead to overall development. Similarly, installation of micro-hydropower projects will provide power for ropeways, a sustainable means of transportation in the fragile landscape of the Himalayas. It is an assumption of this study that diversifying agriculture and enhancing livelihood options sustainably may increase the possibility of food security in the long term.

The food demand of rural households depends on their dual functions, i.e., as households and as firms. As households, their demand depends on income, household size and relative prices, but unlike other consumers, their incomes also depend on changes in relative farm prices because these also change their farm income through crop supply functions. The present exercise limits itself to major food commodities, cereals, cultivation of vegetables and fruits, livestock farming, and timber and non-timber forest products. The per capita domestic consumption of these food groups was derived from the total production adjusted for losses and seed allowances.

Murray (2001) Derives the Following 'Principles' of Livelihoods Research

1. Livelihoods research, by its nature, is essentially carried out at the micro-level: that of 'households' and 'communities'. It involves empirical investigation of combinations of modes of livelihood and, above all, of the relationships between them. It also involves pushing to the limit their potential various methods of understanding changes that have taken place over time.
2. For research into changing livelihoods to be illuminating and useful, however, it is essential to define the structural, historical and institutional elements of what may for convenience be called its macro-context. A time frame must be specified, key variables identified and important trends of change discerned.
3. Insofar as livelihoods research is directed to the diagnosis of the causes of chronic poverty, the circumstances and reasons for poverty should be understood through detailed analysis of social relations in a particular historical context. This implies a structural or relational view of poverty, and, in turn, that understanding of its 'persistence' or intractability or 'deepening' should be driven by questions about inequalities of power.
4. It also implies that livelihoods research and discussion of its implications for 'policy-making' should contain explicit reflection on the particular, relevant, contexts in which 'policy' is made, with reference to key questions.

References

Arimoro, A. O., Fagbeja, M. A., & Eedy, W. (2002). The need and use of geographic information systems (GIS) for environmental impact assessment (EIA) in Africa: With example from ten years experience in Nigeria. *African Journal of Environmental Assessment and Management, 4*(2), 16–27.
Atkinson, E. T. (1889). *The himalayan gazetteer* (Vol. 2, p. 369). Delhi: Cosmo Publication.
Binswanger, H. P., Khandker, S. R., & Rosenzweig, M. R. (1993). 'How infrastructure and financial institutions affect agricultural output and investment in India'. *Journal of Development Economics, 41*(2), 337–366.
Bryceson, D. F. (1999). African rural labour, income diversification and livelihood approaches: A long-term development perspective. *Review of African Political Economy, 26*(80), 171–189.
Chambers, R. (1988). Poverty and livelihoods: whose reality counts? *Environment and Urbanization, 7*, 173.
Chambers, R., & Conway, G. R. (1991). *Sustainable Rural Livelihoods: Practical Concepts for the 21st Century*. Discussion Paper 296. London: Institute of Development Studies.
Conelly, W. T. (1994). Population pressure, labor availability and agricultural disintensification: The decline of farming on Rusinga Island, Kenya. *Human Ecology, 222*, 145–170.
Datt, G., & Ravallion, M. (2002). Is India's economic growth leaving the poor behind? *Journal of Economic Perspectives, 16*(3), 89–108.
Davies, S. (1996). *Adaptable livelihoods: Coping with food insecurity in the malian sahel*. London: Macmillan.

DFID. (2000). *Achieving sustainability, poverty elimination and environment: Strategies for achieving the international development targets*, report from DFID. Retrieved from: http://www.dfid.gov.uk/Documents/publications/tspenvironment.pdf

Diagana, B., & Kelly, V. (1996). *Will the CFA Franc devaluation enhance sustainable agricultural intensification in the Senegalese peanut basin*? USAID and Michigan State University.

Eckholm, E. (1979). *Planting for the future: Forestry for human needs*. World Watch Paper 26, World Watch Institute, Washington.

Eder, J. F. (1991). Agricultural intensification and labor productivity in a Philippine vegetable gardening community: A longitudinal study. *Human Organization, 50*(3), 245–255.

Ellis, F. (1998). Household strategies and rural livelihood diversification. *The Journal of Development Studies, 35*(1), 1–38.

Ellis, F. (2000). *Peasant economics: Farm households and agrarian development* (2nd ed.). UK: Cambridge University Press.

Guojie, C. (2007). Some considerations on strategy of development of mountain regions of China. *Strategy and Policy Decision Research, 22*(2), 126–131. (in Chinese).

Guyer, J. I., & Lambin, E. F. (1993). Land-use in an urban hinterland - ethnography and remote-sensing in the study of African intensification. *American Anthropologist, 95*(4), 839–859.

ICAR. (2002). Indian council of agricultural research, agricultural research data book, ICAR, 2004.

ICIMOD. (1996). *Background Note for Regional Meeting of Experts on Development of Micro Enterprises in Mountain Area* 25–26 July, Unpublished Text.

ICIMOD. (2007). *Background Note for Regional Meeting of Experts on Development of Micro Enterprises in Mountain Area* 25-26 July, Unpublished Text.

ICRA. (1990). 'Intensification and differentiation in *Agricultural and Livestock Production, Usagara Division*, Tanzania'.

Ives, J., & Messerli, E. P. (1989). *The himalayan dilemma: Reconciling development and conservation*. London: Routledge.

Jodha, N. S. (1990). *Mountain perspective and sustainability: A framework for development strategies*. Paper Presented at International Symposium on Strategies for Sustainable Mountain Agriculture, International Centre for Integrated Mountain Development, Kathmandu, Nepal.

Kemp, K., & Martens, P. (2007). Sustainable development: How to manage something that is subjective and never can be achieved? *Sustainability: Science, Practice, and Policy, 3*(2), 5–14.

Khanka, S. S. (1984). Migration from kumaon region: Some findings based on sample study of Pithoragarh district. *Indian Journal of Labour Economics, 26*(4), 302–12

Klaus, J. L. (1983). Rural development in mountainous areas why progress is so difficult. *Mountain Research and Development, 3*(3), 125–129.

Maikhuri, R. K., Saxena, K. G., & Rao, K. S. (1995). Experiences in developing a village agroforestry project in Garhwal Himalaya, India. *International Tree Crops Journal, 8*, 213–221.

Meenakshi, J. V., & Ray, R. (2002). 'How have the disadvantaged fared in India? An analysis of poverty and inequality in the 1990 s'. In K. Sharma (Ed.), *Trade Policy Regim*. UK: Routledge.

Mehta, G. S. (1990). Problems and potentials of horticulture development. In: A. K Singh (Ed.), *Planning strategies for a development region*. Lucknow: Print House.

Mehta, G. S. (1996). *Uttarakhand: Prospects of Development*. New Delhi: Indus Publishing Company.

Murray C. (2001). Livelihoods research: Some conceptual and methodological issues. In: Department of Sociology University of Manchester Background Paper 5 Chronic Poverty Research Centre ISBN Number: 1-904049-05-2

Okali, D. U. U. (1985). The Nigerian environmental; ecological limits of abuse. In: S. Nokoe (Ed.) *Keynotes Paper in the Nigerian Environment; Ecological Limits of Abuse. Proceedings*

Annual Conference and General Meeting of the Ecological Society of Nigeria (ECOSON) (pp. 1–16) May 3rd–5th 1985. Port Harcourt: River state University of Science and Technology
Papola, T. S. (1996). Planning for environment and economic development in mountain areas: Concepts, issues, and approaches. Discussion Paper No. MEI 96/2, Kathmandu, Nepal: ICIMOD.
Ramakrishnan, P. S., Saxena, K. G., Swift, M. J., & Seward, P. D. (Eds.). (1993). *Tropical soil biology and fertility research: South Asian context*. Dehra Dun: Oriental Enterprises.
Reardon, T., Berdegué, J., & Escobar, G. (2001). *Rural nonfarm employment and incomes in Latin America: Overview and policy implications*. World Development (Vol. 29, No. 3). UK: Elsevier Science Ltd.
Roumasset, J. A., Boussard, J.-M., & Singh, I. (1979). *Risk and uncertainty in agricultural development*. New York: Agricultural Development Council.
Sati, V. P. (2004a). Systems of agricultural farming in the Uttaranchal Himalaya, India. *Journal of Mountain Science, 2*(1), 76–85 http://www.imde.ac.cn/journal.
Sati, V. P. (2004b). Systems of vertical Horticulture in the Alaknanda basin of Garhwal Himalaya. *Annals of NAGI, 24*(2), 49–58.
Sati, V. P. (2004c). *Horticultural development in Hills: A case for the Alaknanda Basin* (Ph. D. Thesis). New Delhi: Mittal Publications. www.vedamsbooks.com/38930
Sati, V. P. (2004d). *Uttaranchal: Dilemma of plenties and scarsities*. New Delhi: Mittal Publications. www.vedamsbooks.com/38930
Sati, V. P., & Kumar, K. (2004). *Uttaranchal: Dilemma of plenties and scarsities*. New Delhi: Mittal Publications. www.vedamsbooks.com/38930
Sati, N., & Mansoori, I. K. (2009). *Pilgrimage tourism*. New Delhi: Mittal Publications.
Schelhas, J. (1996). 'Land use choice and change: Intensification and diversification in the Lowland Tropics of Costa Rica'. *Human Organization, 55*(3), 198–306.
Scoones, I. (1998a). Sustainable rural livelihoods a framework for analysis, institute of development studies. *(IDS) Working Paper 72*, Retrieved from www.uvg.edu.gt/instituto/centros/cea/Scoones72.pdf.
Scoones I. (1998b). *Sustainable rural livelihoods: A framework for analysis*. Working Paper 72, Institute for Development Studies, Brighton, UK.
Singh, B. M. (1991). Land capability classification for wasteland management in the Himalayan terrain. In P. N. Gupta & A. K. Roy (Eds.) *Mountain resource management and remote sensing* (pp. 122–135). Dehradun: Surya Publication.
Singh, B. (1998). Vegetable production under protected conditions: Problems and prospects. Indian Society Vegtable Science Souvenir: Silver Jubilee, National Symposium December (pp. 90) 12–14, 1998. Varanasi, U.P. India.
Singh, N., Diwedi, S. K., & Paljor, E. (1999). *Ladakh Mein Sabjion Kei Sanrakshit Kheti*. Regional Research Laboratory of DRDO, Leh. Pub. By D.R.D.O., Leh. Pub. By D.R.D.O. 56 A.P.O.
UNU project on sustainable mountain and forest development. (2002). UNU contribution to the Agenda 21, Chapter 13: Managing Fragile Ecosystems: Sustainable Mountain Development. Retrieved from www.unu.edu/env/mountains
Walker, T. S., & Ryan, J. G. (1990). *Village and household economies in India's semi-arid tropics*. Baltimore: Johns Hopkins University Press.

Chapter 2
Geo-Environmental Setting

Location and Extension

The Himalaya is the youngest and the highest mountain system of the world. It extends about 2400 km as a massive arc, covering an area of about 612021 km^2; from the Pamir's Knot to the Arakan-Yoma mountain ranges of Myanmar and it covers the parts of Afghanistan, Pakistan, India, Tibet (China), Nepal, Bhutan and Myanmar (Fig. 2.1). The Tibetan plateau, known as the roof of the world, forms the northern boundary of this magnificent mountain system while lower extensions of it branch off from eastern and western frontiers of the mountains. Its largest spread is in the Indian Territory and is called the Indian Himalaya Region. It is located in the north part, known as the crown of India. The Himalayan system is very important strategic location—natural, economic and geo-political; it separates the Indian subcontinent from the Central Asia and China. The Himalaya protects the Indian sub-continent from the severe impact of cold waves, blowing from the Tibetan highland. It is believed that if there were no Himalayan ranges, the north part of India would have converted into cold desert. The Himalaya has tremendous economic importance to India as the major rivers originate from it and flow through almost all the northern states of India. The northern plain is nomenclature as 'the Ganges plain', a most fertile, feeds the highest number of population (approximately 50 % population of the Indian sub-continent), and is made of debris deposited by the Himalayan Rivers. Geo-strategically, the Himalaya is relatively important as it protects the Indian subcontinent from the military invasion of the neighbouring countries—China and Pakistan. The Himalaya is divided into three vertical divisions—the Great Himalaya, the Middle Himalaya and the Lesser Himalaya or the Shivalik Ranges; horizontally, it is divided into five divisions—the Jammu and Kashmir Himalaya, the Himachal Himalaya, the Uttarakhand Himalaya, the Nepal Himalaya and the Assam Himalaya (also known as the Eastern Himalaya) (Fig. 2.2). Uttarakhand Himalaya is further divided into the Garhwal and Kumaon Himalayas. The snow clad peaks of the Himalaya present a picturesque and panoramic view and the Himalaya is one amongst the richest areas in terms of biodiversity resources.

V. P. Sati, *Towards Sustainable Livelihoods and Ecosystems in Mountain Regions*,
Environmental Science and Engineering, DOI: 10.1007/978-3-319-03533-8_2,

A case study of the two sub-watersheds—the Kewer Gadhera and Khanda Gad of the Alaknanda Basin was carried out. The Alaknanda Basin represents the eastern part of the Garhwal Himalaya and is extended between 30–31° N and 78° 45′–80° E; it covers an area about 10882 Km^2. Out of the total area of the basin, 433 km^2 is under glacial landscape and rest 288 km^2 is under fluvial landscape. The total number of villages is approximately 2,310. The land under agriculture is 644.22 Km^2, which is 5.9 % of the total geographical area while only 64.8 Km^2 (0.6 %) land is under cultivation of horticultural crops (Sati 2005a, b). This chapter deals with the geography of the Alaknanda Basin that comprises of physiographical divisions, climate and natural resources. Administratively, the four districts of the Garhwal Himalaya i.e., Chamoli, Rudraprayag, Pauri, and Tehri; and a district of the Kumaon Himalaya i.e., Bageshwar comes under the Alaknanda Basin. These districts have eighteen development blocks.

Physiographical Division

The Alaknanda Basin comprises of the hilly terrain, deep gorges and river valley terraces. This basin is broadly divided into four major vertical divisions—the Great Himalayan Ranges, Alpine pastureland, Middle Himalaya and the river valleys. Horizontally, it is divided into the Upper and Lower Alaknanda basins with its different geographical and socio-economic characteristics. Its upper basin covers the part of Bageshwar district, whole Chamoli district and Rudraprayag districts; while lower basin comprises of the parts of Tehri and Pauri districts. The whole basin ranges from about 500 m at Devprayag, where it confluences with the Bhagirathi river to above 8000 m. The physical characteristics of these vertical divisions are as follows:

The Great Himalayan Ranges: The Great Himalayan Ranges are located above 4000 m. Snowline starts at about 3600 m during the summer and at about 2800 m during the winter. The entire region remains snow covered throughout the year. The Alaknanda River and its numerous tributaries originate from the glaciers of the Great Himalayan ranges. The major mountain peaks in this region are Nandadevi (7816 m), Chaukhamba (7138 m), Kamet (7756 m) and Trisul (7120). Nandadevi is the second highest peak of the Indian Himalayan region.

The Alpine Grasslands: Below the snowline and above tree-line, between 2800 m to about 3600 m, alpine grasslands are extensively found. These grasslands are the nature's treasures, providing naturally grown medicinal plants to the local people. The major alpine grasslands are Bedni Bugyal (3354 m), Madhyamaheshwar (3497 m) and Khankharakhet (3100 m). These grasslands remain covered by snow during the winter and in summer, they are major tourists' attraction. The native people of the lowlands immigrate seasonally to these grasslands for summer pasture (Sati 2008a, b, c, d, e, f). The world famous pilgrimages of Badrinath (3100 m) and Kedarnath (3553 m) are located in these meadows.

The Middle Himalaya: This part of the Himalaya comprises of the highlands and mid-altitudes, elevated from 1000 m to about 2800 m. The highlands (2000–2800 m) are mainly covered by the dense forests, including oak and coniferous and remain snow cover during the winter. Mid-altitudes are located from 1000 to 2000 m, are characterised by the high concentration of human population. Agricultural land and human settlements are mostly found in this zone. Many summer resorts are located here where during the summers, climatic conditions remain feasible. These summer resorts play significant role in tourism development of the region. Gwaldam (1875 m), Gopeswar (1293 m), Joshimath (1875 m) and Srinagar (550 m) are the major towns.

Valley Regions: It consists of the river valleys characterised by steep slopes and river terraces; located between about 500 m and 1000 m elevation. The river meeting points, popularly known as *prayags*, are situated in this zone. During summers, the valleys receive high temperature and during winters, the cold waves, which are blowing from the high Himalaya, are present here. Their presence reduces the air temperature. Roads are traversed along the rivers sides and service centres are located on the road heads.

Major Rivers and Their Tributaries

The Alaknanda Basin is source of origin for a number of rivers. Among the major rivers of India, the Alaknanda River and its numerous tributaries (Dauli Ganga, Vishnu Ganga, Nandakini, Pindar, Mandakini, and other perennial streams) make a large river system. Detail descriptions of these rivers are given below:

The Alaknanda River: The Alaknanda River originates from the Alkapuri Bank (Glacier) flows through Badrinath, Vishnuprayag, Nandprayag, Karnprayag, Rudraprayag and finally inlets into the Baghirathi River at Devprayag, where it is called the mother Ganga. Its total length is 195 km and drainage area is 534 crore cubic meter. In the Hindu mythology, it is considered as bridegroom and mother-in-law is the Bhagirathi River. It is also called ‘Vishnu Ganga’ at ‘Badrinath’, a famous pilgrimage devoted to the Lord Vishnu.

The Saraswati River: It is a small tributary of the Alaknanda River, originates from the Great Himalayan Ranges, confluences with the Alaknanda near Mana, a last village of Chamoli district. The meeting point of the two rivers is called ‘Keshavprayag’ which is the first prayag in the Alaknanda River.

The Dhauli River: It outlets from the Niti Pass in Joshimath block (Chamoli District) and inlets into the Alaknanda River at ‘Vishnuprayag’ about 10 km down from Joshimath town towards Badrinath.

The Nandakini River: It outlets from the Nanda Ghunti, flows through the Ghat block (Chamoli District) and inlets in the Alaknanda River at ‘Nandprayag’.

The Pindar River: The Pindar River originates from the Pindari Glacier in Kapkot block of Bageshwar district, flows about 120 km (20 km in Bageshwar

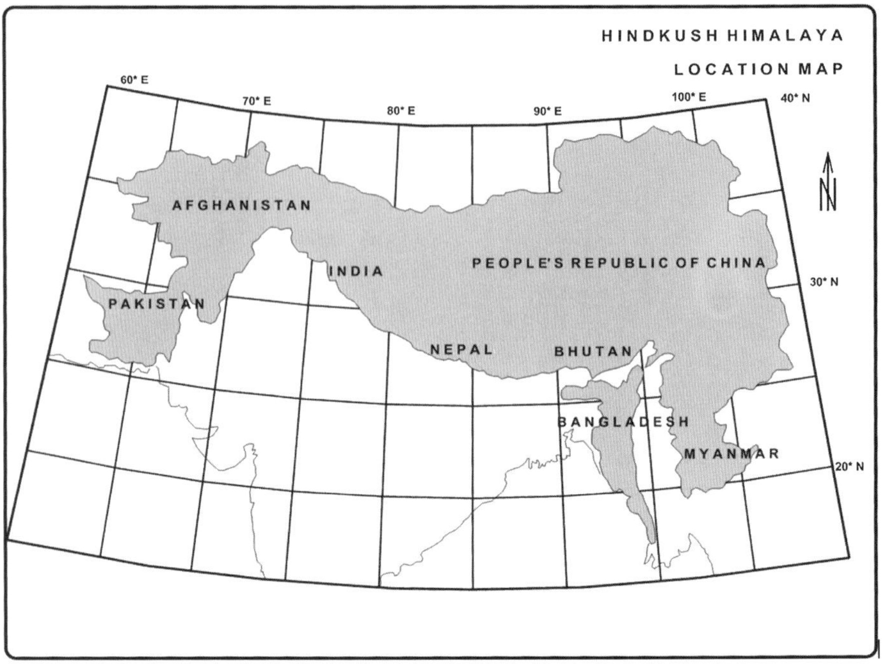

Fig. 2.1 Location map of the Hindkush Himalaya

district and 100 km in Chamoli district), and confluences with the river Alaknanda at 'Karnprayag'.

The Mandakini River: It originates from the Chaura-Bari glacier (Chamoli District) and meets in the river Alaknanda at 'Rudraprayag'. It has numerous small tributaries such as Madhyamaheshwar, Kali and Son rivers (Fig. 2.3).

There are many other perennial streams from small to medium sizes that flow in the Alaknanda River, make this region potentially rich in water resources. Mainly, during the monsoon season, these rivers and perennial streams become very violent; it causes severe damage to agriculture/settlements and to loss of lives.

Climatic Conditions

Climatic conditions are determined by the altitudes, slope aspects, climatic seasons and the presence of the Great Himalayan ranges. The altitudinal differences coupled with varied physiographic divisions, contribute to climatic variations in the Alaknanda Basin. The climate varies from the valley regions (sub-tropical) to the highlands (alpine). Despite diverse physiographic characteristics, sub-regional

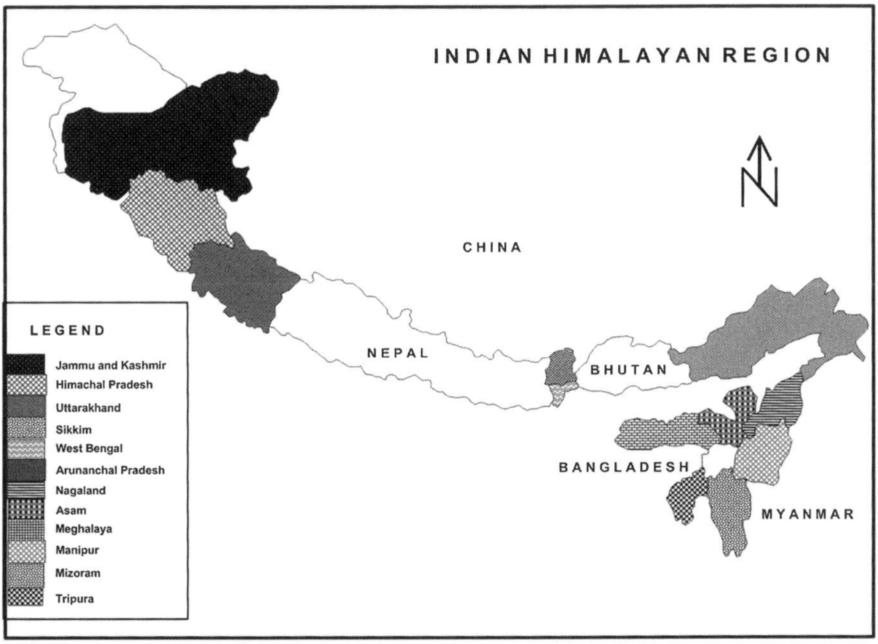

Fig. 2.2 Location map of the Indian Himalayan Region

variations in the average seasonal temperature are not so high. Details of the climatic conditions are presented in the following paragraphs:

Temperature: Temperature varies based on (i) season i.e., summer, monsoon, and winter; (ii) altitudes i.e., from the valley regions (low-lying areas) to the highlands and alpine meadows. The highest temperature was recorded in Srinagar (550 m) in the month of June (30° C) and lowest in Tungnath (3600 m) in the month of Jan (0.5°). Srinagar is a wide valley, covered by the two mountain ranges from the north and the south and between these two ranges, the Alaknanda River flows. It has very contrast climatic conditions. During the summer, it receives high temperature and during the winter, extreme cold waves hit the whole region. As shown in Table 2.1, the whole Alaknanda Basin receives average to low temperature during the entire year. The areas, located above 2000 m, receive heavy snowfall during the winter. Heavy snowfall was observed in Jan 1993, when the low-lying areas (about 900 m elevation), were also covered by snow. Summers are very conducive for health in the mid-altitudes. The valley regions, between Karnprayag and Devprayag, are very warm during summers as monthly temperature averaged about 30° C. During this period, the farming communities migrate to the highlands for summer pasture. Heavy tourist-flow can be observed in the towns of the basin along the Alaknanda River. The large number of pilgrims also visits the two pilgrimages i.e., Badrinath and Kedarnath during the summers.

Table 2.1 Mean monthly temperature in the Alaknanda basin

Name of place	Altitude (m)	Mean monthly temperature ° C											
		Jan	Feb	Mar	Apr	May	Jun	July	Aug	Sep	Oct	Nov	Dec
Srinagar	550	14	18	20	25	25	30	29	28	25	27	17	15
Mastura	1800	4	6	12	14	15	20	20	18	17	14	8	4
Joshimath	1875	2	3	7	11	14	17	18	17	16	10	7	4
Tungnath[a]	3600	0.5	1	3	6	7	12	12	11	5	4	2	1

Sources [a] HAPPRC Srinagar Garhwal (Uttaranchal)
India Meteorological Department, Pune, 1990
Sati (2004a, b, c, d)

Rainfall

Rainfall occurs during the two different seasons—summer and winter. Heavy rainfall occurs during the summer from June to October due to monsoon winds. It is known as orographic rain. The entire region gets intensive rain that varies from the valley regions (low rain) to the highlands (high rain) and from north-facing (leeward) to south-facing (windward) slopes. Due to high and long mountain ranges, leeward and windward phenomenon takes place. Cloudbursts, debris-flows, landslides, landslips, mass movements and flashfloods are very intensive as well as frequent. The entire region is worst affected due to these catastrophes during the monsoon season. During heavy rains, there are many instances where the low-lying river valleys were swept away, causing loss of lives and property. During the winter, rainfall occurs due to western disturbances. The wind that originates through the western disturbances makes the entire region cool and as a consequence, cold waves blow. Mostly during December–January months, precipitation occurs in the forms of snow in the higher reaches above 2000 m and rain in the mid-altitudes and the valley region.

Rainfall data were recorded from the four rainfall data recorded stations of the basin (Table 2.2). These stations are located in different altitudes that ranges from 550 to 3600 m. Highest rainfall was recorded in Okhimath (199.4 cm) followed by Karnprayag (147.1 cm) while lowest rainfall was recorded in Srinagar (92.5 cm). This data reveals that in the high altitude rainfall is higher than the valley regions rainfall. However, there are the regions where low rainfall occurs; for example as Joshimath, which is located at 1875 m, received only 107.5 cm annual rainfall because it is located in the leeward direction.

The cold chilly winters of the highlands and humid monsoon climate in the valley regions characterise the climate. This consequently has effect on the farming system and working potential of the populace. Leeward and windward direction of the slopes determines the amount of rainfall places receives. In this basin, Devprayag is located at 500 m (lowest elevation) while Nandadevi peak (7817 m) is the highest point. The physiographic characteristics also determine the climatic conditions. This basin can be divided into five physiographic zones; the low-lying river valleys, the mid-altitude, the highlands, the alpine pastures and the

Table 2.2 Rainfall data in the major towns of the Alaknanda Basin, 2008

Station	Altitude in m	Annual rain in cm	Seasonal rainfall %			
			Winter	Pre-monsoon	Monsoon	Post-monsoon
Srinagar	550	92.5	16.0	17.7	58.8	8.5
Karnprayag	883	147.1	10.5	13.4	15.9	10.2
Okhimath	1578	199.4	8.8	11.3	71.3	8.6
Pauri	1630	130.3	14.8	14.7	61.5	9.0
Joshimath	1875	107.5	15.4	10.3	53.1	12.2

Source Forest Working Plan, Nainital Working Circle (2010)

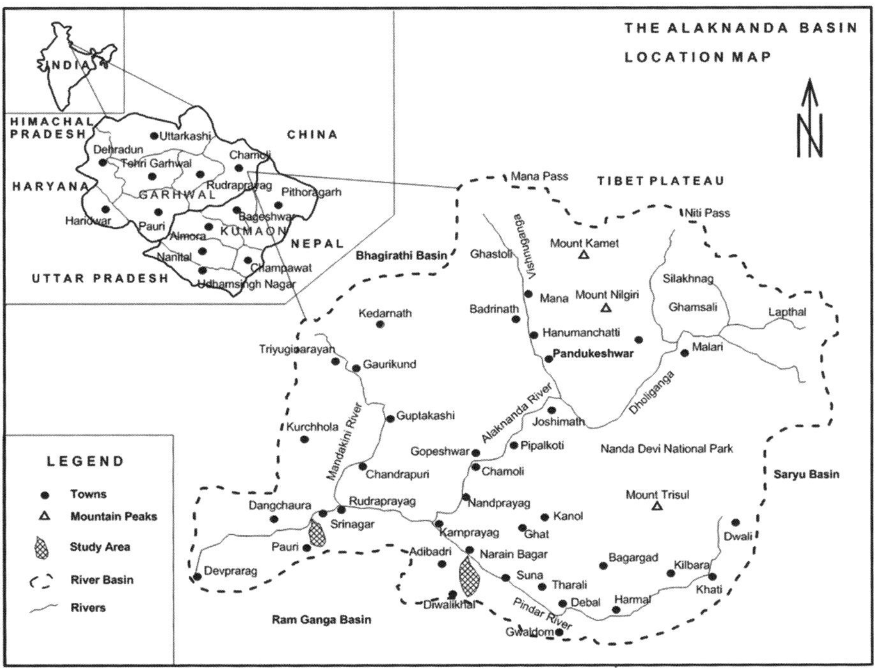

Fig. 2.3 Location map of the Alaknanda Basin showing case study areas

snow-capped mountain peaks. Temperature varies from the valley regions to the highlands and from winter to summer. It reaches down to minus zero degree level in the highlands during the winter. The areas, which are located above 2000 m, receive heavy snowfall during the 4 months of winter. Summer is hot and humid in the valley regions. Meanwhile, mid-altitudes remain mild during the summer and receive heavy downpour. The highland pilgrimages and natural locales are the major attraction for pilgrims and tourists respectively.

Table 2.3 shows rainfall data of the Alaknanda Basin between 1999 and 2004. As the table shows, rainfall intensity and frequency has been changing from year

Table 2.3 Average rainfall from 1999 to 2004

Months	Average rainfall (cm)	1999–2000	2000–2001	2001–2002	2002–2003	2003–2004
May	49	28.9	70.47	93.03	55.87	30.23
June	172.96	135.94	242.65	235	136.36	121.58
July	419.38	314.02	323.99	337.31	183.69	343.48
August	417.41	275.22	374.47	248.84	334.34	347.63
September	201.67	246.6	111.41	40.88	266.9	239.94
October	34.98	39.91	12.97	2.51	10.16	0
November	6.78	0.88	1.36	0.23	3.93	1.8
December	21.92	4.78	13.3	2.56	11.64	13.42
January	51.9	23.54	30.86	30.8	39	45.45
February	59.9	56.15	17.48	39.11	29.69	5.95
March	43.52	24.84	66.82	16.46	16.12	0
April	28.51	28.67	19.09	42.11	30.16	26.33
Total (Average)	125.7	98.23	107.1	90.7	91.1	97.9

Source Compiled by the author from the different sources

to year. It also varies according to (i) the direction of slopes i.e., leeward and windward; (ii) the altitudes i.e., from the lowlands to the highlands. Joshimath, Karnprayag, Okhimath and Srinagar are located in the leeward direction. Therefore, these places receive comparatively less rainfall in comparison to other places that are located in the windward direction as Mandal, Gopeshwar, Gwaldom, Diwalikhal and water dividing regions, facing towards east slopes. Mandal region receives highest rainfall (400 cm) and locally known as Cherrapunji of Garhwal. Average annual rainfall is 125.7 cm. November and December months receive minimum rainfall of 6.78 and 21.92 respectively. During the monsoon period, mainly 2 months of July and August, highest rainfall (above 400 cm) was noticed in the entire region and because of high rainfall, humidity was recorded up to 100 %. The valley regions remain covered by fog during the winter for example, in the months of December-January dense fog occurs in Srinagar Garhwal, a stretch valley of the Alaknanda River and it reduces visibility.

Humidity

This region is characterised by high humidity throughout the year; but the percentage of humidity differs depending upon season and altitude. During the rainy season humidity remains high throughout the whole region whereas, where as, during the summer, humidity decreases especially in the valley regions. Presence of high humidity in the atmosphere provides suitable conditions for growing crops even during the droughts. The drought of 1987, which affected the entire land of India and led to the situations of starvation and food insecurity, did not affect the

Alaknanda Basin (Sati and Kumar 2004). This was predominantly because of the presence of high humidity in the atmosphere. The Great Himalayan Ranges occupy 433 km^2 land of the Alaknanda Basin and the perpetual snow clad ranges of these ranges regulate the climatic conditions viz. temperature, rainfall and moisture.

Climate Change Phenomenon in the Himalaya

Climate change phenomenon is not only a global issue; it also has regional and local dimensions. It has tremendous impact on the land, the oceans and the atmosphere. Meanwhile, mountain regions are worse affected as they are most fragile and vulnerable. This vulnerability is due to the fact that mountains have three dimensional landscapes where high biodiversity—faunal and floral is found. The highly elevated mountain ranges are always covered by perpetual snow. Further, mountains are the major sources of the perennial rivers. Some of the biggest river systems of the world originate from the Himalayan Ranges and more than half of India's population is dependent upon these river systems.

Climate change is observed in the Himalayan region during the past decades. The major changes are seen in the agricultural and horticultural farming along with changing pattern of forest distribution. For example, apples were grown at an altitude of 1600 m during the summer; this is now shifted to the elevation of above 2000 m. This has led to a situation where the cultivation is completely vanished. Similarly, citrus fruits, traditional cereal crops etc., have disappeared from the areas where they were cultivated. As a result, the agricultural land is abandoned.

During the past few decades, simultaneously, changes in the forest distribution pattern were also noticed in the Himalayan Region. The rich diversity of these forest areas is noticeable and it is dominated by the three forest species in the different altitudinal zones. Pine forest lies between 700 and 1600 m, oak forest is found between 1600 and 2200 m and evergreen coniferous forests are located between 2200 and 2800 m. Finally, Alpine meadows are spread above 2800 m. Now, warming of the valleys and the mid-altitudes has resulted in shifting of pine forest to the higher elevation. Consequently, most of the oak forest has been disappeared from their dominating areas (Sati 2006a, b, c, d, e, f).

Potentials of Natural Resources and Their Distribution Pattern

Mountains provide water, biodiversity resources and magnificent landscape. They are the prominent sources of key natural resources—minerals, forests and agricultural products; are storehouse of biological diversity and endangered species;

also caters to great recreational values. As a major ecosystem, representing the complex and interrelated ecology of the planet earth, mountain environments are essential to the survival of the global ecosystem. However, they are rapidly changing. The changes in climate are having its impacts and it is manifested in the form of accelerated soil erosion, landslide, rapid loss of habitat and genetic diversity. As a result, most global mountain areas are experiencing environmental degradation. Further, there is widespread poverty among mountain inhabitants which is accompanied by the loss of their indigenous knowledge. Proper management of mountain resources and socio-economic development of people deserves attention in the wake of global changes.

The Himalayan ecosystem is fragile and diverse. Here, over 51 million people are practicing hill agriculture under high risk and vulnerability. The Himalayan ecosystem is vital to the ecological security of the Indian landmass, as it is providing forest cover, feeding perennial rivers and conserving biodiversity. It also provides a rich base for high value agriculture, and spectacular landscapes for sustainable tourism. It is vulnerable and susceptible to the impacts and consequences of (i) changes on account of natural causes; (ii) climate change resulting from anthropogenic emissions; and (iii) developmental paradigms of the modern society. The Himalaya having largest resources of snow and ice and its glaciers form source of fresh water for the perennial rivers such as the Indus, the Ganga and the Brahmaputra.

Sustainability of the Himalayan ecosystem is crucial for the livelihoods of about 1.3 billion people in Asia. The Himalayan region is rich in biodiversity resources, mineral and power resources, water resources—glaciers and rivers and agricultural resources; yet, high fragility of terrain does not permit to utilise them optimally. Further, any unscientific changes in the ecosystem will lead to ecological degradation. The perennial rivers of north India are largely dependent upon the sustainability of glaciers and the ecosystem of the Himalayan region. In the Himalayan Region, development should be in congruence with the sustainability of the prevailing ecosystem. As mentioned earlier, natural hazards such as terrestrial and atmospheric, are very intensive and frequent throughout the Himalayan region. This restricts and creates hurdles to the development activity.

In the whole world, natural resource management and food security are the prime concerns; this is particularly in the wake of globalization, high growth of population and changing livelihood options. This remains major concern as well as priority for the third world countries where livelihood is dependent on the subsistence agriculture. The Himalayan region is rich in terms of presence of natural resources i.e., flora, fauna and water. Further, rich agro-ecological conditions provide suitability for cultivation of various kinds of crops- subsistence and cash. Water resource has abundance as the major rivers originate from the Himalayan glaciers and flow through this region. These abundant natural resources and suitable agro-climatic conditions are not harnessed sustainably to tackle issues like food insecurity and malnutrition. Lack of infrastructural facilities further accelerates this problem. The economy of the region is largely dependent upon traditional agriculture and on remittances. Meanwhile, the optimum utilisation of these abundant natural resources can enhance livelihoods and the people of the region can have food security. A case study of the natural resources potentials and distribution pattern in the Alaknanda Basin was carried out. The major findings are as follows:

Soil Resources

Soil pattern and texture varies from the Greater Himalaya to the mid-altitudes and the valleys, and accordingly, soil fertility is varied. Landscape of the greater Himalaya is consisted by very steep to steep slopes, are dominantly occupied with very shallow to moderate shallow, excessively drained, sandy-skeletal and loamy skeletal, neutral to slightly acidic, with low available water capacity soils, without profile development in association with rock outcrops. In the Lesser Himalaya, soil can be seen on steep to moderately steep slopes, is shallow to moderately shallow, excessively drained, sandy/loamy-skeletal/loamy with moderate erosion and moderate to strong stoniness. In the side slopes or terrace slopes, soils are moderately deep to deep, excessive drained, fine loamy slightly too moderately acidic with slight to moderate erosion and stoniness. Soils in glacio-fluvial valley consist of moderately shallow excessive drained, coarse loamy, slightly acidic and moderately stony. Fluvial valley's soils are deep well drained, moderately acidic, slightly eroded, and Typic Dystrochrepts.

Water Resources

Water resources of this region remain mostly underutilized; at the same time it is the most abundant resource of the Himalaya. It is estimated that about 11,00,000 million cubic meter water flows every year down the Himalaya which has a potentiality of generating electricity to the tune of 28000 MW; as much as 247000 million cubic meters water is available for irrigation in the Indo-Gangetic plains (Valdiya 1985). Per capita fresh water availability in the Himalayan Region is calculated to range from 1757 m^3/yr in Indus, 1473 m^3/yr in Ganges, 18417 m^3/yr in Brahmaputra with an all India average of 2214 m^3/yr. The Alaknanda Basin is endowed with bounty of water resources accounting for about 8 % of the total water resources in the country. Unfortunately, this vast potential has not been rationally exploited. Endowed with huge water resources potential, it has also the worst water resource problems rendering untold sufferings to millions every year. The region experiences excessive rainfall and high floods during monsoon months and also suffers from acute shortage of drinking water in many areas due to lack of management. The Alaknanda river and its numerous tributaries (like Dauli Ganga, Vishnu Ganga, Nandakini, Pindar, Mandakini and their sub tributaries) together, which are perennial and glacial fed, presents huge water resources reservoir (Sati 2006a). In many areas, the tributaries

Table 2.4 Water resource potential in the Ganges system

Name	Origin	Confluences	Length in km	Annual drainage (Crore cubic m)
Tons	Har-Ki-Dun	Dakpathar	148	484.4
Yamuna	Yamunotri	Dhalipur	136	165.1
Bhagirathi	Gaukukh	Devprayag	205	253.3
Alaknanda	Alkapuri bank	Devprayag	195	534.2
Nayar	Dudhatoli	Vyasghat	87	162.6
Kosi	Kausani	Sultanpur Patti	168	187.0
Saryu	Tungbhadra	Pancheshwar	146	135.0
Ram Ganga	Dudhatoli	Kalagarh	155	97.2
Kali	Lipulekh	Tanakpur	252	238.7

Source Primary data collection with the help of 'Survey of India Toposheeds'

and sub-tributaries provide ideal sites for micro-hydropower projects. The area of unlimited water resources is facing acute water shortage of drinking and irrigation facilities; sustainable utilisation of water through construction of micro-hydropower projects would definitely solve the duo problems (Sati 2008a). The basic issues underlying the water resources problems are: recurring floods, drainage congestion, soil erosion, human influence on environment etc., and this call for its integrated management so that water requirement for drinking, irrigation, hydropower and recreation can be ensured.

Water, a joint product of land and forest, is the third largest resource of the Himalayan region. The major rivers of India i.e., the Ganga, the Yamuna, the Saryu, and the Kali along with their numerous tributaries, originate and flow from this region. There are 238 glaciers in the Central Himalayan region. Water from these perennial sources can meet the requirements of the whole country, provided it is properly managed. The river Ganga itself can meet the 42 % of the national water need according to estimation. It flows through the eight states of India and meets the demands of tremendous water requirement. Meanwhile, the Central Himalayan region faces severe water crises through out the year. Here, about three thousand villages have acute water scarcity. The Ganga River irrigates only 4 % land. It is estimated that this region receives 663 billion KL water every year from rain. This plenty of water can be used for drinking, irrigation and industrial purposes. Furthermore, if water potential is to be linked with other industries such as packed water, cold water, tourism (water sports) and electricity generation, and if a considerable amount of tax to be paid to the government by the lowlands users, it can become a major resource and can change the economic scenario. There is a potential of generating 30,000 MW electricity. Until now, only 3000 MW electricity is being generated and it serves the lowland areas mostly (Table 2.4).

Management of water resource is a crucial issue because of the undulating terrain and fragility of landscape, which does not permit for construction of macro-level dams. Developmental interventions at micro-level considering drinking water, irrigation and hydroelectricity generation in an integrated manner, have yet to be properly designed and tested. Traditional management of water resource as a form of *gharat* (water mills) and *gools* (small canals) did not involve any advance technology, while they are absolutely appropriate for this ecologically fragile mountain terrain (Sati 2006b).

Forest Resources

Among the other natural resources in the Alaknanda Basin, forests are most important, both economically and environmentally. The geographical area covered by forest is reported to be 10,21,156 ha, which accounts for around 42.2 %. Ownership of the forest in the state is mainly shared between the forest department (69.1) and Civil and Soyam (community forest (23.4). Forest Panchayats (6.9) and private forests (including cantonment) are other two stake holders and they manage the remaining area. The alpine, temperate and sub-temperate forests that cover most parts of the basin are the natural habitats for some of the best-known wildlife creatures. Alpine forests in the region include Valley of Flowers National Park (known for its amazing variety of flowers), Nanda Devi National Park and Govind Ghat National Park.

The Alaknanda Basin is very rich in terms of forest and its diversity. Right from the valley region to the highly elevated Alpine meadows, locally known as *Kharak* or *Bugyal*, the diversity in plants is very rich and extensive. In the middle altitude, Pine (*Chir*) are found while in the upper reaches, temperate coniferous forest mainly *Kharsu* (*Quercus semicarpifolia*). *Tilonj* (*Q. dilitata*), *Rianj* (*Q. lanuginose*) and *Banj* oak (*Q. leucotricophora*) are abundantly found (Sati 2006a, b, c, d, e, f). Except these forest types, many other fodder plants like, *bhimal* and *khadik*, are also grown along with edges of agricultural fields. The main forests are (i) Deodar Forests (*Cedrus deodara*) are found between the heights of 1650 and 2300 m; (ii) Blue Pine Forests (*Pinus wallichiana*) are also known as *Kail* [These are found in Joshimath areas mostly mixed with Deodar Forests. The tree occurs between 1650 and 2300 m. Timber is used in making sturdy cupboards and pelmets in houses]; (iii) *Chir* Forests (*Pinus roxburghii*) are found in the whole region [Its forests exist mostly between the altitudes of 1000 and 1650 m. It is used for making packing cases and paneling in interior decoration. It is also used as fuel wood]; (iv) Oak Forests (*Quercus species*) [found in the basin between the heights of 1325 and 1625 m. It is used for fuel wood and charcoal manufacturing. It is the best firewood having high caloric value. It is a broad-leaved tree]; (v) Fir (*Abies pindrow*) and Spruce (*Picea smithiana*) Forests are found mostly between 2300 and 2950 m.

Altitude regulates diversity in flora in the Alaknanda Basin. According to altitudinal zones, various kinds of flora with great economic value are found. Most of the forests belts in the basin are inaccessible. Consequently, their economic use is just negligible. In the high altitude, these forests help to increase the soil fertility because soil is brought with rainwater and deposited in the lowlands. The entire basin is ecologically fragile. Landslides and landslips are very common, particularly during the rainy seasons. Due to heavy rains (known as cloudburst) and steep slopes, this situation is further accentuated. Forests are the main tools for conserving soil and land. To conserve the soil and land, diversity in flora is required.

Diversity in flora is found in all altitudinal zones, dominated by oak and pine forests. Forest covers about 42.2 % of geographical land and it is increasing constantly. A study on land cover change shows that about 1.3 % forest cover increased in the last three decades (Sati 2008b). Forest is the main source of livelihood of the populace. It provides fodder, firewood, timber, non-timber products, herbs and environmental services. Altitudinal variations in forest types are due to changes in the climatic conditions. This region comprises of subtropical to temperate, alpine and cold climatic zones resulted in diversity in natural vegetation gradually from the valley regions to the uplands. Oak and pine forests are useful for firewood, fodder, and timber. The farming community of the region is highly dependent on forests for their livelihood. Pine forests are found mostly in the valley regions and the mid-altitudes patches while oak forests have monopoly over the highlands. The impact of global change can be noticed here as pine trees invaded oak in many areas.

The Himalayan region is amongst the 12 biodiversity hotspots of the world and it has forests ranging from sub-tropical to temperate to alpine. Here, the economic viability of the forest resources is tremendously high therefore, the populace of the region has been engaging in collection of timber and not-timber forest products. In the Indian Central Himalayan Region (ICHR), forest covers above 65 % land area. The economy of the region is based upon the cultivation of traditional cereal crops whose production and per ha yield is considerably low. Thus, the people largely depend on the forest resource for fodder, firewood, and food. Forests have the linkages with agriculture and crops production. For manure composition, most of the tree leaves are used to mix-up with cow-dung that enhances productivity. In the Himalayan region, oak and pine are largely used for firewood and construction of building. Oak leaves are also used as important fodder for the lactating animals. It enhances milk production thus, food security can be obtained. There are numerous forest products which are used as spices to food and medicinal plants and herbs for traditional health care system. The others are essential oils, fibers and silk, natural dyes and organic products, and bees and bee products. Wild fruits as *kafal, hensole, kilmode, bhamore* and many others substantially enhance livelihood options on which the local people are dependent. This illustration reveals that forests are the major source of livelihood thus, need attention for a comprehensive conservation measures. The traditional methods of harnessing non-timber forest products are sustainable that can be restored for the future use of forests.

In the Himalayan region, there are 2,300 guldar, 240 tigers, 1,350 elephant, 250 kasturi mrig, 10,800 sambhar, 10,500 kakad, 5,000 giddha, 53,000 chital and 400 bird's species found in twelve national parks and wildlife sanctuaries. Valley of Flowers, Asan Barrage, Nanda Devi and Gangotri National Parks are the main attraction. The revenue from the forest was Rs. 9150 lakh during 2001–2002 and 20316 lakh in 2007–2008 (Sati 2013). The attitude of the previous governments and colonial rulers towards forest conservation and utilisation of its products was not very positive. Currently, the forest department is also following the same pathway. Local people have been raising the issues related to their forest rights and are agitating against the Forest Act (FA). The FA is a main hindrance to the construction of national highways, power lines, irrigation, and drinking water projects and the Act even prevents the establishment of schools and colleges. Around 200 development projects are pending due to the FA. The local people are unable to utilise forest resources let it be- timber, fodder or other forest products. From forestation to prevention of fire, the forest department has failed to ensure local people participation. This region has privilege of having some of the Worlds' pioneering Research Institutes such as the Wild Life Institute, the Forest Research Institute, the Forest Survey of India, the Botanical Survey of India, the Zoological Survey of India, the Wadia Institute of Himalayan Geology, the Survey of India, the GBPIHED, the Indian Institute of Petroleum, and the Oil and Natural Gas Commission. But, these institutes could not do much in livelihood enhancement of the local people mostly because of lack of coordination at different levels and between different stakeholders. Forest fire is a major manmade disaster here. In the month of May 2009, there were about 1400 incidences of forest fire. These incidences had greater intensity than the historical forest fire of 1921 and 1995. Forest fire spread about 3000 ha land, killed eight people, and injured two dozen people. Due to implementation of the Wild Life Act (WLA), number of wild animals has shown a remarkable increase, and consequently, man-animal conflict and human deaths also has shown increase. About 200 people died and more than 500 people were injured from the last 9 years (Times of India 2009).

Land Resource and Environmental Services

Land management is a crucial issue in the Himalaya region. Only 8 % land is cultivable with high population pressure and cropping intensity. It is also used for diverse activities. During pre-independence period there were inter-relationship between land, forest and the people. The local people were fully dependent on land, animal and forest. Forest products were the major source of livelihood. In pre-independent India, agricultural land was around 20 %. First FA came into being with classification of forests in 1863. This led forest inaccessible to the local people and put peasants into severe trouble. In 1923, another FA came into force. Some amendments were made after rigorous opposition of the local people. Until independence, the montane people squeezed out to rehabilitate in the productive areas of Doon, Dwar, Tarai, and Bhabar. This was further accelerated by the UP Jamidari Emancipation Act of 1966. Under this Act, community land was converted into forestland. As a result of this, agricultural land could not get any support for extension even after getting the status of a separate state. The government announced to increase agricultural land up to 20 % but it could not do so.

Shrinking agriculture land and mounting population pressure put agricultural land unproductive. It was resulted in a large-scale emigration.

Himalaya Mountain regulates the climatic conditions. Forest sequestrates carbon. Since Uttarakhand Himalaya has abundance of forests, it helps to reduce carbon in the atmosphere. The issues of *Jal, Jangle*, and *Jameen* got attention to provide services to the sustenance of the local people. For the well being of local people, FA and WLA should be amended. Rights should be given to the local people to run their livelihoods from forest and its products. In all activity, participation of the local people should be ensured. Taxes should be imposed on water which runoffs to the plain region. Eco-tourism and cultivating medicinal plants with involvement of the local people will enhance livelihood. Unless agriculture could get a base for livelihood, out migration will be continued. There is need to re-think about the agriculture. Either the per capita land should be increased or the optimum utilisation of the arable land should be ensured using scientific innovation. The compensation for all these environmental service should be given to the people.

Livestock Rearing

Livestock plays foremost role in the livelihood as it is the second main occupation after farming of subsistence crops. On the other, it helps agriculture systems as plowing the field and providing manure. For centuries, organic fertilizer is used as manure, for production of crops. Besides, production of milk plays substantial role to run livelihood. Milk is mostly consumed domestically and is seldom sold in the nearby service centres. Draught animal constitutes the composition of domestic animal kingdom mostly uses in the field of agriculture as draught power. Cows and oxen are outnumbered followed by buffaloes. In the upland areas, goats are reared and used for wool and meat. Recent study on livestock farming reveals that the numbers of domesticated animals are decreasing (Sati and Singh, 2010). The households earlier had more than five cattle now have one or two each. However, the changes in rearing livestock vary from the valleys to the highlands. The valleys are connected by motorable roads. The households have changed their occupation from agriculture to tertiary sector. Consequently, the numbers of livestock reduced considerably. Contrary to this, in the upland areas, subsistence farming system is the main stay, which is based upon the mixed agriculture-livestock farming. Livelihood options have also changed in the uplands as outward migration is increasing and this creates a situation where the major source of income is from remittances. As a result of this, the agricultural practices and livestock are showing retardation.

The climate and agro-ecological conditions provide a base for rearing of high yield variety (HYV) of animals as temperate climate and alpine meadows are widespread in the region. During the 1980s, veterinary centres where established in each block headquarter and HYV lactating cows were given to the farming communities. Nevertheless, this scheme could not meet with any success. A study carried out by the present author that this was due to mismanagement. The government distributed HYV animals to the poor farmers who were unable to manage them. *Murrah* buffaloes, reared in the upland areas, are high lactating animal. Meanwhile, their numbers are less as compared to draught animals. Rearing of *murrah* buffaloes and high lactating cows can contribute substantial household income. Livestock rearing constitutes major segment in the income and economy of the farming community. It received more significance, when high attention was paid to organic farming during recently. Keeping suitability of landscape, climate and agro-ecological conditions in mind, rearing of draught animals, and lactating cows and buffaloes for enhancing income and economy is inevitable.

Potential of Environmental Services

Water (*Jal*), forest (*Jungle*), and land (*Jameen*) are the three major life sustaining components in the ICHR. About 70 % population of the region are fully dependent on these components for their livelihoods. By and large, the benefits of ecosystem services to the people of the highlands are minimal as they do not receive any payments for these resources (mainly from the inhabitants of the lowland areas, who are noticeably benefitted). For the last many decades there has been debates and expectation that the people of the montane areas of ICHR will get their rights in terms of payment of ecosystem services. Such payment was to be made by the people of the lowland areas; but until now nothing has materialised and no policy is formulated. The natural and ecosystem services of the state can increase the annual turnover more than the annual turnover of the corporate companies and it has the capacity of providing sustainable livelihoods to the local people satisfactorily. It is the need of the hour that the issue of payment of ecosystem services should be raised at the national level so that the populace of the ICHR can enjoy their livelihood rights. The ICHR is the abode of rivers and forest resources. River water is still untapped and it directly runs-off to the Ganges plain, where it irrigates large agricultural land. The construction of micro-dams on these rivers will provide electricity and water not only to the local areas but also to the states of northern India.

References

Nainital Working Circle. (2010). Unpublished record of the Department of Forest, Government of Uttarakhand.

Sati, V. P. & Kumar, K. (2004). *Uttaranchal: Dilemma of plenties and scarsities*. New Delhi: Mittal Publications. www.vedamsbooks.com/38930

Sati, V. P., & Singh, R. B. (2010). Prospects of livestock farming in the Uttarakhand Himalaya, India. *Journal of Livestock Science*, *1*(1), 9–16. http://livestockscience.in/wp-content/uploads/2010/08/sati

Sati, V. P. (2004a). Systems of agricultural farming in the Uttaranchal Himalaya, India. *Journal of Mountain Science*, *2*(1), 76–85 http://www.imde.ac.cn/journal

Sati, V. P. (2004b). Systems of vertical horticulture in the Alaknanda Basin of Garhwal Himalaya. *ANNALS of NAGI*, *24*(2) 49–58.

Sati, V. P. (2004c). *Horticultural development in hills: A case for the Alaknanda Basin,* New Delhi: Mittal Publications, (Ph.D. Thesis) www.vedamsbooks.com/38930

Sati, V. P. (2004d). *Uttaranchal: Dilemma of plenties and scarsities*, New Delhi: Mittal Publications. www.vedamsbooks.com/38930

Sati, V. P. (2005a). Vertical zonation of horticultural farming in the Alaknanda Basin of Garhwal Himalaya, India. *Journal of Mountain Science*, *2*(4), 319–328. http://www.imde.ac.cn/journal

Sati, V. P. (2005b). Natural resource conditions and economic development in the Uttaranchal Himalaya, India. *Journal of Mountain Science*, *2*(4), 336–350. http://www.imde.ac.cn/journal

Sati, V. P. (2006a). Forest resource management in mountain regions: A case for the Pindar Basin of Uttaranchal Himalaya. *Lyonia: A Journal of Ecology and Application*, *11*(1), 75–84 www.lyonia.org

Sati, V. P. (2006b). Role of off-season vegetables in the sustainable livelihood of hills' people—A case for the Pindar Basin (Uttaranchal Himalaya), published in the proceedings of International symposium Towards Sustainable Livelihoods and Ecosystems in Mountainous Regions 7–9 March 2006, Chiang Mai, Thailand Web Link: https://www.uni-hohenheim.de/sfb564/uplands2006/contributions/PS-A-Vishwambhar-full.pdf

Sati, V. P. (2006c). Population and sustainability issues in mountains: A case for the Uttaranchal Himalaya. *ENVIS on Himalayan Ecology,* 14(2). http://gbpihed.nic.in/envis/HTML/vol14_2/vpsati.htm

Sati, V. P. (2006d). Natural hazards in an ecologically mountain Terrain: A case for the Pindar Basin of Uttarancal Himalaya. *ENVIS on Himalayan Ecology, 14*(1). http://gbpihed.nic.in/envis/HTML/vol14_1/vpsati.htm

Sati, V. P. (2006e). Tourism practices and approaches for its development in the Uttaranchal Himalaya, India. Published electronically in *Mountain Forum Online Library*. http://www.mtnforum.org/oldocs/528.pdf

Sati, V. P. (2006f). Water resource potentials and management in the Pindar Basin. Included in Chapter 4 of 2006 Rosenberg Water Policy forum: Preliminary Public E-consultation. A focused e-discussion held from July 31–August 28, 2006. http://www.banffcentre.ca/mountainculture/mtnforum/conferences/rosenberg/week2.asp

Sati, V. P. (2008a). Traditional Intramontane mobility in Garhwal Himalaya: A survey of subsistence practices in the Pindar Basin, Uttaranchal. *Singapore Journal of Tropical Asia*, *29*(2), 173–185. A peer reviewed journal of Blackwell Publications. www.blackwell/journal/sjtg_v29_i2/sjtg_333.

Sati, V. P. (2008b). Farming systems and strategies for sustainable livelihood in Eritrea. *African Journal of Food Agriculture Nutrition and Development*, *8*(2), 219–237. A peer reviewed quarterly journal. www.ajfand.net/issue19/PDFs/Sati2505.pdf

Sati, V. P. (2008c). *Natural resource management and food security in the Alaknanda Basin of Garhwal Himalaya, India. Peer reviewed proceedings of 'Tropentag 2008' an International Conference, Germany*. Stuttgart, Germany.

Sati, V. P. (2008d). Natural resource management and food security in the Garhwal Himalaya. *ENVIS on Himalayan Ecology, 16*(2), 6–16. http://gbpihed.nic.in/envis/HTML/vol16(2)/vpsati.htm

Sati, V. P. (2008e). *Natural resources management and sustainable development in Pindar Valley, Himalayas.* Dehradun: Bishen Singh Mahendrapal Singh.

Sati, V. P. (2008f). People's participation in forest resource management in the Uttaranchal Himalaya. *MF Bulletin*, *8*(2), July 2008. www.mtnforum.org

Sati, V. P. (2013). Cultivation of medicinal plants and its contribution to livelihood enhancement in the central Himalayan region. *Advancement of Medicinal Plant Research, Net Journal, 1*(2), 17–23.

Times of India. (2009). Daily English newspaper published from Delhi, Report on Wildlife Act of India, Sep 30, 2009.

Valdiya, K. S. (1985). Himalayan tragedy. Big dams, seismicity, erosion and drying-up of springs in the Himalayan region. *Central Himalayan Environment Association Bulletin, 1*, 1–24.

Chapter 3
Socio-Economy and Population Profile

Introduction

Mountain regions—worldwide—are economically backward and environmentally degraded. In this region, the level of economic development is not sufficient to bring the living standard of the populace to a desired level. Though the natural resources are abundantly found in the forms of water, flora and fauna they are not harnessed optimally. The same is the case with abundant human resources, and as a result, the mountain regions remain economically backward. Environmental and ecosystem degradation has been seen to have negative consequences beyond their margins (Eckholm 1975). However, on-going research indicates that some consequences have not been fully or adequately substantiated (Ives and Messerli 1989).

ICHR, also known as the Uttarakhand Himalaya, is an integral part of the Himalayan Mountain ranges. It is amongst the richest regions in terms of biodiversity; at the same time it is also amongst the economically backward regions of the country. Economic development has not taken any shape and direction in this region. The economy rests on the cultivation of traditional cereal farming, while the scope for cultivation of cash crops like off-season vegetables, fruits, medicinal plants and utilisation of non-timber forest products remains tremendously high. This region is characterised by highly elevated mountains and steep precipitous slopes. These steep mountainous areas have their own characteristics. Cloudburst triggered debris flow and flashfloods are very common. Mass movements and landslides are common features which can be observed along the roads or on the course of river valleys. Land resources like forests, water, mineral, soil and physical landforms have deteriorated considerably due to excessive erosion and frequent occurrence of mass movement. The heterogeneity in climate and high anomalies in the landforms ensures diversity in the potentiality and capability of crops.

Trends in sectoral and sub-sectoral growth in economy since 1993–2004 show two distinct kinds of growth dynamics. There are six sub-sectors that exhibit steady and unchanging growth dynamics for the whole period. These are (i) Agriculture;

V. P. Sati, *Towards Sustainable Livelihoods and Ecosystems in Mountain Regions*, Environmental Science and Engineering, DOI: 10.1007/978-3-319-03533-8_3,

(ii) Forestry; (iii) Fishery; (iv) Mining; (v) Real Estate; Ownership of Dwellings & Business Services and (vi) Banking and Insurance. On the other hand, the remaining seven sub-sectors show a distinct jump in growth rates since the region attained Statehood in 2000. These are (i) Manufacturing; (ii) Construction; (iii) Electricity, Gas and Water; (iv) Trade, Hotels and Restaurants; (v) Transport, Storage and Communications; (vi) Other Services and (vii) Public Administration. These high rates of growth are partly due to the small base of these sectors; while other factors like better administration and governance, together with fiscal incentives for private participation in these sectors, have also played an important role. Some of these factors will remain significant even in the long run, while others will have a temporary impact on the growth rates for a short period of time. Clearly, the capacity to sustain high growth rates will depend on whether the long-run factors remain more relevant than the temporary factors.

Sustainable development has been used widely as an organising framework since the World Conservation Strategy, the Brundtland Report (WCED 1987) and UNCED. The idea of sustainable development, as promoted in the World Conservation Strategy, refers to a broad range of objectives for meeting basic human needs, while maintaining essential ecological processes and life-support systems, preserving genetic diversity and ensuring sustainable utilisation of species and ecosystems. The term sustainable development gained popularity and scholarly attention with the establishment of the World Commission on Environment and Development (WCED). The commission's report was long on problem descriptions and short on policy prescriptions, but it provided the standard definition for sustainable development as "development that meets the needs of the present without compromising the ability of future generations to meet their needs" (WCED 1987).

Economic Development

The ICHR has been recognised as one of the backward and underdeveloped regions of India since the very beginning of the planning era. But the initial planning efforts did not reflect this concern/recognition and, in fact, the policy planners did not feel the need for any area-specific development approach. Instead, both the plains and hills of Uttar Pradesh state were treated as parts of a single framework of development. During the Third Plan period, some consideration was given to the development of the ICHR by providing a separate allocation of Rs. 500 million for specific activities. During the early 1970s, a decision was made by the Central Government to treat Uttarakhand as a special region for development planning purposes. A department for Hill Area Development was created and assigned the task of identifying areas for potential development. Further, evaluation and monitoring of these development programmes were also the responsibility of the newly established department. Since the Fifth Five-Year Plan (1974–1979), separate development plans have been prepared for the ICHR and the Central Government is providing about 50 % in assistance to the state budget for the development programmes. On principle, it has been agreed that central assistance to the development plans of the ICHR would be at par with assistance being given to Himachal Pradesh.

The plan outlay for the ICHR has been increasing consistently during different five-year plan periods. The provision for a separate budget for this region was introduced in the Fifth Five-Year Plan period (1974–1979) with a plan outlay of Rs. 20.40 billion. It increased to Rs. 65.8 billion during the Sixth Five-Year Plan period (1980–1985), to Rs 121.31 billion during the Seventh Five-Year Plan period (1985–1990) and to Rs. 210.5 billion during the Eighth Plan (1992–1997). Similarly, the central assistance has increased from Rs. 10.40 billion in the Five-Year Plan period to Rs. 67.92 billion in the Seventh Five-Year Plan period. In the Eighth Five-Year Plan, it was further increased to Rs. 100.5 billion. However, if we quote the figures in terms of percentages, then the central assistance to the total plan expenditure declined from 50.98 % during the Fifth Five-Year Plan period to 47.74 % during the Eighth Five-Year plan period.

The absence of financial resources and funds is not the major hurdle, but it is the absence of a region-specific approach that has resulted in relatively poor and unsustainable development of the region. In the initial years, the development strategy as reflected in the sectoral priorities and programme plans for the region was no different from the overall strategy adopted for the whole state, which predominantly has a topography of plains. With the Seventh Five-Year Plan (1985–1990), the constraints and potentials of the region were recognised by planners and policymakers to some extent. As a result, the development priorities as reflected in public sector plan outlays placed greater stress, for example, on soil and water conservation, including watershed management, water supply, sanitation and minor irrigation works. Other sectors like road transport and basic education were also given renewed and improved priority.

The Eighth Five-Year Plan (1992–1997), while continuing the focus on these priorities, gave a more focused consideration to the issues related to environmentally sustainable development. The plan document emphasised afforestation, scientific management and protection of forest; integrated soil and water conservation; and watershed development. The plan also focused on creating additional employment and income opportunities through diversification of agricultural activities like horticulture, encouragement of pollution-free and locally available raw materials-based small-scale and cottage industries. The development of the tourism sector was also emphasised upon.

Despite the fact that most relevant aspects find place in the plans, particularly from 1985 onwards, the stated goals of development have remained unfulfilled. Consequently, unsustainable pattern of development and poverty has persisted and inequality in income distribution has shown a rising trend. The number of job seekers on the live registers of employment exchanges rose from about 89,000 in 1981 to 2,43,171 in 1991. In the 7 years from 1984 to 1991, employment in the organised public and private sectors increased only from 2.07 lakh to 2.36 lakh (Dhar 1996). After 2000, Uttarakhand's average GSDP is 10.28. Table 3.1 shows growth rate of GSDP in Uttarakhand from 2000–2001 to 2005–2006 (DES 2010).

Table 3.1 Growth rate of GSDP in Uttarakhand

No.	Year	Growth rate of GSDP at constant (1999–2000) prices
1	2000–2001	11.83
2	2001–2002	5.48
3	2002–2003	9.52
4	2003–2004	7.74
5	2004–2005	7.78
6	2005–2006	10.28

Source Directorate of economics and statistics, Uttarakhand (2010)

Infrastructural Facilities

Infrastructural facilities in the region are developing. In 2005, density of rail routes was 6.45 route km for every 1,000 sq km. The region developed 345 km of rail routes by 2005–2006. The state envisages to posit air transport as a reliable all-weather transport option for the hills. Upgradation work at the Pantnagar Airport to facilitate international flights is under progress. The Jolly Grant Airport at Dehradun is amongst the best airports in India. The government has allocated US$ 0.23 million for construction of at least one helipad per district. The total length of the roads in the state is 27,509.58 km. Of a total of 15,656 villages in the state, 9,419 have proper road connectivity.

Some of the country's best institutes of higher learning are also located in this region. GB Pant University of Agriculture and Technology in Pantnagar; Indian Institute of Technology in Roorkee; Forest Research Institute in Dehradun; GBPIHED in Almora; Rural Biotechnology Complex in Kosi and Herbal Research and Development Institute, Gopeswar are some of them. With a current literacy rate of 79.63 %, (Census of India 2011), the state is striving to achieve 100 % literacy rate. The state plans to achieve 100 % computer literacy among the student population. For this purpose, the state has taken initiatives to improve the IT infrastructure at all levels; and the initiatives has begun with educational institutions.

The state has transformed from a predominantly agrarian to a hub of industrial activity. During 2005–2006, the state successfully developed three Integrated Industrial Estates (IIEs) at Haridwar, Pantnagar and Sitarganj. Pharma City at Salequi, Information Technology Park at Sahastradhara (Dehradun) and Growth Center at Siggadi (Kotdwar) are other industrial establishments. Uttarakhand also saw the development of 20 Private/Joint/Co-operative sector industrial sectors in Public–Private Partnership mode. Some of the other efforts made in this direction include: (1) Provision of broadband connectivity for all industries in the state; (2) Proposed strengthening of single window—contact, information and facilitation mechanism during 2006–2007.

The mountain economy is predominantly agriculture centric; people also depend upon remittances (money order economy). There are three groups of people. The first group belongs to families that have permanently emigrated and settled elsewhere. They have left their land abandoned. The second group comprises families with one or more members working in the tertiary sector and other members working in agricultural fields. The last group constitutes those people working only in agricultural fields. This group is the most deprived group in terms of earning income and struggle for their livelihoods; this group represents the largest proportion of society.

Sustainable livelihood is the prime concern, particularly in the wake of globalisation, high growth of population and changing livelihood options. This is prominent in the Himalayan region, where livelihood is largely dependent on the cultivation of subsistence cereal crops. The Himalayas are very rich in terms of presence of natural resources, i.e. flora, fauna, magnificent landforms and water. Further, rich agro-climatic conditions are substantially widespread, useful for cultivating various kinds of crops—subsistence and cash crops. Water resources are abundantly found in the major rivers though these resources are underutilised. This existing situation has resulted in food insecurity and malnutrition, which is a common and growing phenomenon. Lacking in infrastructural facilities—transportation, banking, marketing, education and institutions, further accelerates the difficulties.

Perspectives of Natural Resource Management

The ICHR has feasible climatic conditions and it is a reservoir of many natural resources. Abundance in these natural resources can be harnessed in a sustainable manner for the overall development of the region. Sati and Kumar (2004) have suggested several steps towards optimal use of natural resources and they are as follows: (1) Horticultural development offers significant scope and could become a leading sector in many parts, particularly on the mid-slopes and water dividing of the rivers and their tributaries. In certain areas, however, the fruit growing belts are well developed and substantial quantities and varieties of fruit are supplied to the market through contractors and other intermediaries. Increasing participation of the farming communities in horticulture would also require the provision of food security through the public distribution system to farmers who are shifting their agricultural land into fruit farming. (2) A large variety of minor forest produce, medicinal plants, natural dyers and natural fibre plants are plentiful; these resources could be harnessed advantageously to produce useful and high-value items. (3) Promotion of ringal (small species of bamboo) plantation in larger uncultivable areas for making useful marketable products could thus be both economically feasible and environmentally sustainable. Marketing arrangements and training to develop new skills for making diversified products are the measures needed. (4) Rearing sheep and goats has special advantages in the mountains at altitudes of 700 m and above. It has been traditional in the Bhotia households in high areas. They have been using the wool from sheep and goats to produce carpets, shawls and sweaters. Now, besides the Bhotia Community, various other institutions, private entrepreneurs and government departments are engaged in the production of woollen materials. The Animal Husbandry Department of the State Government is involved in improving sheep breeds so as to develop high-quality woollen materials. Improved varieties of sheep are supplied to local sheep breeders for cross-breeding to achieve this. (5) Woollen industries, including rearing goats and sheep, have been given significant importance in the past. A greater section of the population, particularly women and relatively low-income groups, is engaged in the production of woollen items as a part-time or full-time activity. This has led to an increase in the demand for raw wool and scarcity of supplies has become a major constraint due to the expansion of this activity. Improvement in supplies requires action on two fronts; one, improvements in the breeding and nutrition of sheep to increase wool yields and, two, organised procurement of wool from outside (including from other countries). (6) Tea gardens are present in this region for centuries, but now their numbers have declined considerably. Following the creation of Uttarakhand state, the government has launched a scheme for establishing tea gardens in different geographically ideal locations. There are many suitable belts for tea production; among these at many sites tea gardens are being established and at many places they have to be re-established. (7) Harvesting water resources in the form of hydroelectricity generation and water supply for drinking and irrigation purposes will surely lead to sustainable development of the region. (8) Cultivation of herbs is suitable in the region because of extensive alpine meadows where varieties of herbs are grown naturally. (9) Cultivation of off-season vegetables is the best way to reduce food scarcity, malnutrition, poverty and to enhance the livelihood options of the local populace. The diversity in the climate is also a major conducive factor. Thus, the role of climate is noteworthy. Here, the production of potato, onion and tomato has shown an increase in recent times and it has reached a production level where potato is a commercial crop and has entered the regional markets. Other vegetables and spices are pumpkin, cucumber, beans, garlic, ginger, turmeric, radish, carrot and green leaves but these are grown at a domestic level, although their commercial viability is high and they flourish in all parts of the region. (10) Promotion of dairy farming is possible due to plenty of grassland, diversity in climate and availability of market. These are the major factors which determine the promotion of dairy farming. HYV animals can be reared, which will help in high production of milk and will contribute to the regional development. (11) High biodiversity makes the state an attractive destination for Agro and Food Processing Industries. The state offers immense opportunities for establishing agro and food parks. (12) Rare species of plants and animals found in this region add to its natural advantage in this sector. The government plans to develop Uttarakhand as a centre of excellence in biotechnology and proposes to grant industry status to units coming up in the sector; provisions made for the IT sector shall be applicable to this sector as well.

The regional development perspective emphasises more upon environmental and economic issues and thus social issues are placed at the periphery. Environmental degradation is a product of negative relationship between people to people rather than people to habitat (Horowitz 1988). Development priorities for the mountains are justified for improving the quality of life of mountain people and also to persuade the people of the plains that the future of the mountains cannot be isolated from their own future (Eckholm 1975). However, to what extent a more marginal area like the Himalayan region will get priority over the Indo-Gangetic plains will continue to be a question that has to be resolved by the political forces. Protection of interests of the indigenous population, therefore, must assume the focus as well as priority for interventions aimed at sustainable development in the Himalayan region. Unfortunately, human dimensions in environmental and developmental changes remain poorly understood (Fisher 1990).

Growing concern over the deteriorating environment, which is the outcome of activities by different stakeholders over the last four decades seems to have links with the gigantic cause-and-effect arguments on which the Himalayas and the northern plains are at the receiving end are also plunging towards environmental and socio-economic collapse. The processes—physical, human, socio-economic and political have contributed to such debates through visions of deforestation, landslides, large-scale downstream flooding, uncontrolled population growth, increasing poverty and malnutrition. This pattern of thinking has been widely accepted as an established fact by a large number of people who often lend their support to perpetuate it as a truism. An appraisal of the literature available on the Himalayan region suggests that management consideration, argued to have emanated from scientific studies, is inherent in the traditional systems. Thus, what is needed is to appreciate the positive points of traditional resource production and use systems that strengthen them through science and technology input for further improvement in their values and efficiency, instead of advocating abrupt changes involving replacement of traditional systems with new ones found suitable elsewhere but whose applicability in the mountain systems such as Himalaya is not clearly established. The most important development objectives in this context are: (i) improving food and energy supplies by increasing agricultural and forestry products with appropriate modification of the production structures; (ii) securing the highest possible employment and income for the broad mass of the population and (iii) long-term protection of natural resources. One may observe that these objectives are interdependent in the long term and too much emphasis on any one of them would not provide self-dependence between natural, technical and social factors with respect to the goal attainment needed. Since ecosystems are the natural basis for all development processes and their maintenance is of predominant importance in the longer term, planning process, therefore, should take this into consideration.

Neither the lack of general awareness about the problems of hill areas nor the lack of sincerity on the part of the Central/State Planners seems to be the basic problem with development planning. The problem lies in lacking an integrated and region-specific approach to planning and implementation of these development programmes. Thus, environmental protection issues were approached without placing due regard to the needs of the people; infrastructure was developed without much attention being paid to its effective use for local development and sectoral economic activities were ‘promoted’ through incentives and subsidies without simultaneously promoting the development of activity-specific infrastructure/services and inter-sectoral linkages. Adequate recognition was not given to the fact that linkages necessary to the development processes do not materialise on their own, nor even easily, in hill and mountain regions, as they can be expected in the plains. Diversity in mountain areas demands a highly decentralised, area-based approach. The approach has to be different not only from that of the plains but also inter-region differences have to be appreciated and taken into consideration.

Considering the past experiences of the unsatisfactory impact of development programmes, it is necessary to modify planning strategies and approaches after considering the geographical, economic and socio-cultural conditions of the region. Formulation of such an area-specific planning model for the region is not practical from either the physical or administrative angles because the region comprises a small part of India in terms of population and geographical area. This absence of an area-specific approach to development planning seems to be the main issue for discontent in the region, and it has given rise to a movement for the formation of a hill state. Leaving aside political interpretations of the movement, campaigners for a separate state argue that a region-specific approach to development will be possible for Uttarakhand given its special geographical characteristics, only when the region has a reasonable degree of autonomy to plan for its own development; mere decentralization is not adequate. What needs to be done is to divide the State into planning regions identifying specific characters of resource based, nature, and magnitude of problem, planning strategies, required public interventions needed specific financial allocation, as well as anticipated target. These planning regions could be any one or combination of the following (Sati and Kumar 2004):

(A) Intensive Development Regions having inputs (foothill belt, Doon, Bhabar)

1. Dairy, horticulture and consumer industries development
2. Short-duration tourism, pilgrimage and professional education development

(B) Extensive Development Regions with moderate inputs (Inner and Lesser Himalayas)

1. Long-duration tourism, horticulture, cottage industry development
2. Small-scale hydropower-afforestation-horticulture-cold storage
3. Tourism-animal husbandry-vegetable development-cold storage

(C) Extensive Development Regions with heavy inputs

1. Long-duration tourism-pilgrimage-hydropower development
2. Adventure tourism-pilgrimage-horticulture-pastoralism development
3. Hydropower-adventure tourism-cottage industry-animal husbandry development

(D) Moderate Development Regions

1. Agriculture-horticulture-agro-industries
2. Agriculture-small-scale industries-fruit processing development.

The planning process of the region should be based on the strategy of the state's plan to be worked out on the basis of the area-specific plans of individual planning regions. Minor adjustments could be done at later stages to maintain the growth targets of the state.

Population Profile of the ICHR

In the ICHR, population distribution, density of population, sex ratio, literacy and decadal growth rate are all influenced by pull and push factors such as landscape, nature as well as gradient of terrain. Infrastructural and transportation facilities also influence the population dynamics. It was noticed by the author during rapid field visits to the study area that high concentration of population is found in the mid-altitudes where small patches of terraced agricultural fields are located. The valleys are narrow and steep and the highlands are inaccessible and characterised by harsh climatic conditions; therefore, population distribution is sparse and population density is low. High literacy rate, high level of education and minimum opportunities of employment compel people to emigrate from the mainland to the fertile plains of Tarai, Bhabar, Doon and Dwar. Therefore, the growth rate of population and population density varies from the mainland (low) to the fertile plains (high). The male population is known as white colour workers and generally they do not work in agricultural fields. Women are considered as the backbone of the economy because most of the household and agriculture work is managed by them. Men are either employed in the armed forces or are working in schools as teachers.

Population data of 2001 and 2011 of the ICHR were collected from the Census of India (COI) and analysed to illustrate sex ratio, literacy rate, population density and decadal population growth. Table 3.2 shows district-wise population of the ICHR in 2001 and 2011. The total population in the entire ICHR region was 1,01,16,752 in 2011, while it was 84,89,349 in 2001. Hardwar district registered the highest population of 19,27,029 in 2011 and 14,47,187 in 2001. It was followed by Dehradun district from 12,82,143 in 2001 to 16,98,560 in 2011. Udhamsingh Nagar ranks third (16,48,367 in 2011 and 12,35,614 in 2001). Rudraprayag district registered the lowest population followed by Champawat and Bageshwar districts. Haridwar, Dehradun and Udhamsingh Nagar districts are located in the fertile plain, therefore, concentration of population is higher in these districts. Districts like Rudraprayag, Champawat and Bageswhar are small in area and are located in the remote areas where infrastructural and transportation facilities are unavailable and the population size is considerably small.

Table 3.2 District-wise population of ICHR

District	Population 2001			Population 2011		
	Total	Male	Female	Total	Male	Female
Uttarkashi	295,013	152,016	142,997	329,686	168,335	161,351
Chamoli	370,359	183,745	186,614	391,114	193,572	197,542
Rudraprayag	227,439	107,535	119,904	236,857	111,747	125,110
Tehri	604,747	295,168	309,579	616,409	296,604	319,805
Dehradun	1,282,143	679,583	602,560	1,698,560	893,222	805,338
Pauri	697,078	331,061	366,017	686,527	326,406	360,121
Pithoragarh	462,289	227,615	234,674	485,993	240,427	245,566
Bageshwar	247,163	117,374	129,789	259,840	124,121	135,719
Almora	632,866	294,984	337,882	621,927	290,414	331,513
Champawat	224,542	111,084	113,458	259,315	130,881	128,434
Nainital	762,909	400,254	362,655	955,128	494,115	461,013
Udhamsingh Nagar	1,235,614	649,484	586,130	1,648,367	858,906	789,461
Haridwar	1,447,187	776,021	671,166	1,927,029	1,025,428	901,601
Uttarakhand	**8,489,349**	**4,325,924**	**4,163,425**	**10,116,752**	**5,154,178**	**4,962,574**

Source Census of India (2011)

Table 3.3 Decadal growth rate of population

District	Decadal population growth	
	1991–2001	2001–2011
Uttarkashi	23.07	11.75
Chamoli	13.87	5.60
Rudraprayag	13.43	4.14
Tehri	16.24	1.93
Dehradun	25.00	32.48
Pauri	3.91	−1.51
Pithoragarh	10.95	5.13
Bageshwar	9.28	5.13
Almora	3.67	−1.73
Champawat	17.60	15.49
Nainital	32.72	25.20
Udhamsingh Nagar	33.60	33.40
Haridwar	28.70	33.16
Uttarakhand	**20.41**	**19.17**

Source Census of India (2011)

Table 3.3 presents district-wise decadal growth rate of population in the ICHR. Except Dehradun and Haridwar districts, where decadal growth rate increased from 1991–2001 to 2001–2011, all other districts of Uttarakhand registered less growth rate; Almora district even registered a negative growth rate during the same period. In Udhamsingh Nagar, decadal growth rate of population decreased in small proportions. The high decadal growth rate in Dehradun and Haridwar districts during 1991–2001 and 2001–2011 was due to high immigration in these areas.

Table 3.4 Sex ratio

District	Sex ratio	
	2001	2011
Uttarkashi	941	959
Chamoli	1,016	1,021
Rudraprayag	1,115	1,120
Tehri	1,049	1,078
Dehradun	887	902
Pauri	1,106	1,103
Pithoragarh	1,031	1,021
Bageshwar	1,106	1,093
Almora	1,145	1,142
Champawat	1,021	981
Nainital	906	933
Udhamsingh Nagar	902	919
Haridwar	865	879
Uttarakhand	**962**	**963**

Source Census of India (2011)

The people of the remote and mountainous districts emigrated largely to Dehradun and Haridwar districts after Dehradun became the capital of Uttarakhand in 2000. In terms of the decadal growth rate in 2001–2011, Uttarakhand has higher growth rate (19.17) in comparison to the national average (17.64).

The state of Uttarakhand is among the states of India where the ratio of women per 1,000 men is high compared to the national average of 940 ((Table 3.4). According to COI 2001 it is 963 women per 1,000 men. In some districts, the sex ratio is above 1,100. These districts are Rudraprayag, Pauri, Bageshwar and Almora. Other districts like Chamoli, Tehri, Pithoragarh and Champawat also registered a sex ratio above 1,000. Sex ratio has been decreased in Pauri and Champawat districts from 2001 to 2011. In some districts such as Dehradun, Nainital, Udhamsingh Nagar and Haridwar, the number of women per 1,000 men is quite low. Rural areas recorded better sex ratio than urban areas. Dehradun, Nainital, Udhamsingh Nagar and Haridwar districts have high urban population, therefore, sex ratio is comparatively low in these districts.

When compared with other Himalayan States the population density in the state of Uttarakhand is the highest (189 in 2011), except in Tripura (where proportion of mountain region is less). But when compared to the national figure (384), it is about half of it. It varies from the remote and mountainous districts to districts that are located in the fertile plain areas which are called Tarai, Bhabar, Doon and Dwar. Table 3.5 shows that the highest density of population is found in Haridwar district (817) followed by Udhamsingh Nagar (648) and Dehradun (550) (Census of India 2011). There is remarkable increase in the density of population in these districts from 2001 to 2011. By contrast, Uttarkashi, Chamoli and Pithoragarh have very low population density, i.e. 41, 49 and 69, respectively. Low population density in these districts is generally due to steep and precipitous terrain, less sown area and

Table 3.5 Population density

District	Population density	
	2001	2012
Uttarkashi	37	41
Chamoli	46	49
Rudraprayag	115	119
Tehri	166	169
Dehradun	415	550
Pauri	131	129
Pithoragarh	65	69
Bageshwar	110	116
Almora	201	198
Champawat	127	147
Nainital	179	225
Udhamsingh Nagar	486	648
Haridwar	613	817
Uttarakhand	**159**	**189**

Source Census of India (2011)

poor infrastructural facilities. These push factors led to the exodus emigration of population to the plain areas. The author observs that there are many villages, particularly in the highlands, where about one-third of the population emigrated during the last two decades.

The state of Uttarakhand has a high literacy rate (79.63 %) in comparison to the national average (74.04 %). Among the districts of Uttarakhand, Dehradun has the highest literacy rate (85.24 %) followed by Nainital, where the literacy rate is 84.85 %. Udhamsingh Nagar (74.44 %) and Nainital (74.62 %) have the lowest literacy rates. As shown in Table 3.6, the male literacy rate is higher than the female literacy rate in all districts.

Table 3.7 illustrates the decadal population growth of Uttarakhand state from 1901 to 2011. The highest population growth rate was observed in the 1970s which was 27.45, followed by the 1960s (24.42). In the 1980s, population growth rate was 23.13. It shows that from the 1960s to the 1980s decadal population growth rate was high. Negative decadal population growth rate was observed in the decade of 1911–1921 which was -1.23. It was followed by the 1900s (8.20) and 1920s (8.74). In 1901, total population of Uttarakhand (then called Uttar Pradesh Hills) was 19,79,866 which by 2011 increased to 1,01,16,752.

Social Composition

The social composition of the Himalayan region is woven by the various fabrics of culture and customs. Followers of all religions inhabit this region. The famous highland pilgrimages are the major attraction for pilgrims. Devotees from world

Table 3.6 Literacy rate

District	Literacy rate (2011)		
	Total	Male	Female
Uttarkashi	75.98	89.26	62.23
Chamoli	83.48	94.18	73.20
Rudraprayag	82.09	94.97	70.94
Tehri	75.10	89.91	61.77
Dehradun	85.24	90.32	79.61
Pauri	82.59	93.18	73.26
Pithoragarh	82.93	93.45	72.97
Bageshwar	80.69	93.20	69.59
Almora	81.06	93.57	70.44
Champawat	80.73	92.65	68.81
Nainital	84.85	91.09	78.21
Udhamsingh Nagar	74.44	82.48	65.73
Haridwar	74.62	82.26	65.96
Uttarakhand	**79.63**	**88.33**	**70.70**

Source Census of India (2011)

Table 3.7 Decadal population growth, 1901–2011

Year	Population	Population growth
1901	1,979,866	–
1911	2,142,258	+8.20
1921	2,115,984	−1.23
1931	2,301,019	+8.74
1941	2,614,540	+13.63
1951	2,945,929	+12.67
1961	3,610,938	+22.57
1971	4,492,724	+24.42
1981	5,725,972	+27.45
1991	7,050,634	+23.13
2001	8,489,349	+20.41
2011	10,116,752	+19.17

Source Census of India (2011)

over and from the Indian sub-continent visit these highland pilgrimages every year. The role of highland pilgrimages is noteworthy for cultural transformation, because pilgrims and tourists visit these areas and leave their cultural footprints. The waves of modern civilisation can be seen in the major routes leading to these pilgrimages. Tourism has also transformed the cultural structure of the mountain regions.

The society is divided into four *vernas* (castes)—Brahmin, Shatriya, Vaishya and Shudra. They are followers of Lord Shiva. The sacrificial system prevails, where people sacrifice goats to Shakti, the Goddess of power. Fairs and festivals are celebrated throughout the year with folk dances having nature's rhythm.

The people are peace lovers and hard-working; they worship nature in different forms—trees, animals and mountains. Culture and customs are their way of life. The mountain people are highly influenced by global changes and the impact of these global changes on them is high. This high impact can be attributed to their innocence and earlier isolations.

References

Census of India. (2011). India's Census report published by the registrar publication, Delhi.

Dhar, T. N. (1996). *Proceedings and recommendations of the SHERPA seminar on fodder problems in the Himalayan region of India, Pashulok, Rishikesh, India, 22–23 Dec 1997.*

Eckholm, E. P. (1975). The deterioration of mountain environments. *Science, 189*, 764–770.

Fisher, R. J. (1990). The Himalayan dilemma: Finding the human face. *Pacific View Point, 31*, 69–71.

Horowitz, M. M. (1988). Anthropology and the new development agendas. *Development Anthropology Network, 6*, 1–4.

Ives, J., & Messerli, E. P. (1989). *The Himalayan dilemma: Reconciling development and conservation*. London: Routledge.

Sati, V. P. & Kumar, K. (2004). *Uttaranchal: Dilemma of plenties and scarsities*. New Delhi: Mittal Publications. www.vedamsbooks.com/38930

WCED. (1987). *Our common future*. World commission on environment and development. Oxford: Oxford University Press.

Chapter 4
Sustainable Livelihoods: Diversifications and Enhancements

Attainment of sustainable livelihoods in mountain regions is inevitable and is the need of the hour. The geographical conditions of the region are very diverse and harsh in terms of landscape vulnerability and terrain structure. Undulating terrain, low fertility of soils, inadequate irrigational facilities, minimum scope for modernisation of agriculture and as a result low production and per ha yield of crops are the other major impediments towards sustainable livelihoods. The traditional subsistence cereal farming further accelerates this problem, with respect to the maintenance of food security. Meanwhile, the scope of cultivating cash crops other than biomass-based agriculture has tremendous potential to enhance the livelihood to the level of self-sufficiency. This chapter focuses upon two aspects, viz. (i) the potentials and viability of cash crops—off-season vegetables, fruit cultivation, dairy farming, tea cultivation; and (ii) the use of non-timber products—resin, wild fruits, cultivation of medicinal plants, bee—beekeeping and natural dyes to maintain sustainable livelihood and to attend food security in this fragile mountain terrain.

Cash Crops

Prospects of Off-season Vegetables

Introduction

India is the second largest producer of vegetables in the world. At the same time, it has not achieved self-sufficiency in this area. Efforts are directed to increase the production of vegetable crops at the national level. The report of the Indian Council of Agricultural Research (ICAR 2002) envisages that the present vegetable production of 90.8 million tonnes is to be raised to 250 million tonnes by 2024–2025. Governmental efforts are noteworthy to produce the substantial amount of vegetables so that the target could be achieved. The major efforts are bringing in additional areas under vegetable crops and also promoting the use of

V. P. Sati, *Towards Sustainable Livelihoods and Ecosystems in Mountain Regions*, Environmental Science and Engineering, DOI: 10.1007/978-3-319-03533-8_4,

hybrid seeds and improved agro-techniques. Another effort and potential approach is perfection and promotion of protected cultivation of vegetables (Singh 1998; Singh et al. 1999).

The Himalayan mountain system, due to its altitudinal and geographical diversities, has the potential to cultivating all varieties of vegetables. Most of the areas produce very high quality of vegetables; at the same time, when we take the favourable agro-ecological conditions into consideration it can be concluded that the production potentialities are under-utilized. The villages located in the highlands export potatoes to the regional market and earn a high income. Meanwhile, the other parts of the region—the mid-slopes and the valley regions—produce vegetables for domestic consumption only. Production and per ha yield of vegetables can be increased tremendously with little effort and government support because a suitable agro-climatic environment is already existing.

The Alaknanda basin constitutes an integral part of Uttarakhand Himalaya. It is characterised by rough, rugged and undulating terrain. Agricultural farming is carried out mainly on the narrow patches of terraced fields. Subsistence cereal farming dominates the farming systems, which is quite insufficient for livelihood of the people of this region. As a result, widespread poverty and malnutrition are very common features of this region. The struggle for managing livelihood smoothly is higher than in any other region of the Himalayas. The consequence of this is the high rate of out-migration to the Himalayan foothills and Ganges plains. Also, the terrain does not permit intensive cultivation of agricultural crops mainly due to the presence of steep slopes, high rate of soil erosion, instability and fragility of the landscape. Under such circumstances, cultivation of off-season vegetables can help in maintaining the sustainable livelihoods of the rural poor people and can provide more employment to the unemployed youth of the region.

Diversity in the agro-ecological conditions provides suitability for growing a variety of vegetables on the mountain slopes. High production of these crops can considerably assist in poverty alleviation and manifest to reducing malnutrition. In the Alaknanda basin, the production of potato, onion and tomato is noteworthy. Other vegetables that grow throughout the year are pumpkin, cucumber, beans, radish, carrot, coriander and greens, but these are grown in less proportions and are consumed domestically. Meanwhile, the commercial viability of these vegetables is quite high and the geo-climatic conditions support the mass production of these crops. Among spices ginger, turmeric, chili, *tejpat* (a medicinal plant used as a spice), coriander and garlic are largely grown. Recently, among other vegetable crops, the production of potato has attained impressive production figures, mostly in the highlands (1500–2200 m). Onion production is also sufficient in the valley regions and in lower middle altitudes (800–1200 m).

Conditions for Sustainable Farming of Off-season Vegetables

The Alaknanda basin provides a wide range of agro-ecological conditions for growing various kinds of off-season vegetables of high quality and quantity.

Suitability of agro-ecological conditions and the low return from cereal crops further enhance a base for cultivating off-season vegetables. Detailed discussions on the prospects of sustainable off-season vegetables farming are as follows:

Suitable Agro-Ecological Conditions

Agro-ecological conditions are extremely suitable for cultivating off-season vegetables in the Alaknanda Basin, normally in all altitudinal zones—the valley regions, mid-altitudes and the highlands. The nature of terrain, slope aspects, soil contents and availability of water are favourable for off-season vegetable cultivation. In the mid-altitude and highlands, potato is grown widely, while the terraced valley regions are suitable for producing onion. These crops have very high economic viability and grow in the two different altitudes.

Potato Farming in the Highlands

Potato is the main vegetable crop in the highlands. It is grown between 1500 and 2200 m, mostly on gentle slopes (10–15°). During the 1980s, the farmers of the region started intensive potato cultivation. Slowly, it moved to large-scale production and export. Along with this, other off-season vegetables and spices such as beans, ginger, cucumber, pumpkin, turmeric and chilli are also grown and locally consumed.

Onion Farming in the Valley Regions and Mid-Altitudes

Onion is grown in the valley regions where availability of irrigation is ample. It is grown during the summer in different localities along the river terraces between 800 and 1200 m. It is both locally consumed and exported to the local market. Besides these crops, this region has tremendous scope of growing spices and green vegetables. These crops may support sustainable livelihoods provided they are produced commercially.

Low Return from Traditional Cereal Farming

Traditional cereal farming is the main occupation of the people of the Himalayan region in general and the Alaknanda basin in particular (Sati 2005a, b). The output from traditionally grown crops is not sufficient as even these crops do not meet the daily food requirements. The nature of terrain, slope aspects, soil infertility, unavailability of water and no use of modern innovations in the field of agriculture are the main constraints for food security and sustainable development of the region. This situation has led to shifting from cereal crops to cultivating off-season vegetables, to a certain extent. Table 4.1 shows the production (quintal) and productivity (per ha yield) of traditional crops and vegetables in 2003.

The author conducted a case study of six villages located at different elevations, viz. from 750 to 2,200 m (Table 4.1). A comparative study of the traditional crops (millets, wheat and rice) and vegetables (onion and potato) was done. The table presents area, production and productivity of the traditional crops and vegetables. The data show that the production and productivity of vegetables is comparatively higher than cereal crops, both in the valleys and in the highlands. Meanwhile, the

Table 4.1 Production (quintal) and productivity (per ha yield) of traditional crops and vegetables (2003)

Name of case study village	Elevation	Traditional crops (millets, wheat and rice)			Vegetables (potato and onion)		
		Area devoted (ha)	Production	Productivity	Area devoted (ha)	Production	Productivity
Dimri	750	465	1,600 (Rice and wheat)	4.4	55	400 (Onion and other seasonal vegetables)	7.2
Kulsari	1150	460	1,700 (Rice and wheat)	4.7	85	600 (Onion and other seasonal vegetables)	7.1
Kewer	1200	510	1,850 (Rice and wheat)	3.6	65	500 (Onion and other seasonal vegetables)	7.6
Lolti	1800	370	1,005 (Millets)	2.7	136	1,600 (Potato and other seasonal vegetables)	11.7
Khainoli	1900	385	970 (Millets)	2.5	165	1,745 (Potato and other seasonal vegetables)	10.5
Kwarad	2200	380	945 (Millets)	2.5	180	1,800 (Potato and other seasonal vegetable)	10

Source Primary collection

land under cereal crops is high. In the highlands, land under vegetable crops is comparatively higher than in the lowlands. An example is taken from Table 4.1 that compares size of land and productivity of crops for cereals and vegetables. About 460 ha agricultural land was devoted for cereal crops in Kulsari village (1150 m) and its productivity was 4.7, whereas land under onion was only 85 ha and its productivity was 7.1. By contrast, the land under traditional crops in the highland village of Kwarad (2200 m) was 380 ha and its productivity was 2.5. Meanwhile, about 180 ha land was devoted for vegetable crops and productivity was 10.

Promotion of Off-season Vegetable Farming

As already mentioned in the previous paragraphs, the Himalayan region possesses suitable agro-ecological conditions for cultivating off-season vegetables therefore, the high varieties of onion, ginger, garlic, capsicum, cauliflower, ladies' finger, cucumber, pumpkin, tomato and potato can be produced at commercial level. It was observed that the land under vegetable crops is proportionally less. Along with some important steps taken by governmental agencies towards innovation of farming systems, the farmers of the mountain regions transformed their agricultural land from cultivating cereals to cultivating cash crops. This step was taken mainly because of the high return from cash crops compared to traditional cereal crops. However, the land put under cash crop was remarkably less.

Changing Pattern of Cereal Crops

Cultivation of cereal crops dominates in the farming systems of the Himalayan region. It is a centuries-old practice and is the main occupation of the people upon which their livelihoods are dependent. Crop types and intensity vary depending upon altitudes and seasons. Generally, millets are grown in the highlands while paddy and wheat are grown in the valleys and mid-altitudes. The influence of the monsoons on the cropping pattern is dominant, with the result that the total cropped area of about 70.75 % is under '*Kharif*' or rainy season crops. In the region, whatever the type of soil or amount of rainfall, the dominance of foodgrains in the cropping pattern is obvious everywhere (Sati and Rawat 1993). Economic viability of these food crops is insufficient, in terms of meeting food requirements of the populace. As a result, the farmers turned towards cultivating off-season vegetables and they shifted their agricultural land to cultivating off-season vegetables. This change in the land was minimal and so does not require a dominant space in the study. Further, high growth in population, its pressure on the cultivable land, and modern innovation in the field of agriculture have all together led to increase in the cultivation of cash crops for securing a livelihood. However, technological innovations such as chemical fertilisers, pesticides and high yield varieties of crops got a tremendous response in the valleys, but could not change the highland farming system to a very great extent, due to mountain specificities. Forests and livestock provide material and energy inputs in traditional mountain farming systems, thus expansion of traditional agriculture runs the risk of forest degradation. In order to meet the present and future challenges meeting sustainability criteria, traditional systems need to be adapted in ways which enhance crop yields but not at the environmental and social costs (Ramakrishnan et al. 1993).

Presently, the cultivation of foodgrain is limited only to the terraced slopes of the mid-altitudes (below 2000 m) or unterraced gentle slopes of the highlands (above 2000 m). In terms of agro-biodiversity, it is quite high in the highlands compared to the valleys. Crop diversity is managed by mixed cropping and crop rotation (Sati 1993).

Discussion and Conclusions

Cultivation of cereal crops dominates the cropping pattern in the Himalayan region though the economic viability is considerably less. The main reasons are (i) traditional methods of cultivation; (ii) low quality seeds and poor land condition; (iii) and infertile soil. Due to these factors, crops production as well as productivity is below the optimum level. Fragmentation of agricultural fields and fragility of terrain further accentuated these impediments and they do not support the poor farmers to cultivate farmlands frequently and intensively. Modern innovations in agriculture cannot be used properly and sufficiently. This is primarily due to undulating terrain and steep slopes. The mode of agriculture is traditional and entirely dependent on draught power. The use of chemical fertilizers is not feasible. Past experience has shown that the fertility of the soil has lessened due to overuse of chemical fertilisers in the 1980s. Because agriculture is rain-fed, chemical fertilisers reduce soil fertility under this condition. Debates exist among the stakeholders about the viability of subsistence cereal crops, though the debate is still inconclusive as no consensus has been reached. However, some conclusions have been drawn from the studies. These results show that cereal crops are very suitable to the ecosystems of mountain regions and they are also economically unviable. Meanwhile, the agro-ecological conditions are tremendously high for cultivating cash crops, especially off-season vegetables. In some areas, it is cultivated but the pace is slow and land under its cultivation is less. If more land is devoted to off-season vegetables, food security can be ensured. Scientific training programmes can be imparted to farmers to encourage them to adopt cash crop farming. Infrastructural facilities such as construction of cold storages, availability of markets, improved seeds and adequate manure are required to enhance the production of cash crops (Table 4.2).

Fruit Cultivation

The Himalayan region presents a wide and unique opportunity for cultivation of a variety of fruits, citrus and apple in particular, as the agro-ecological conditions in this region are quite suitable. However, this region could not position itself well in terms of production and productivity of fruits. At the same time some of the

Table 4.2 Cultivated vegetable crops and availability time

Local name	English name	Botanical name	Availability season
Lauki	Bottle gourd	*Lagenaria vulgaris* L.	Rainy
Gol Caddu	Pumpkin	*Cucurbita maxima* Duchesne	Rainy
Tit Karaila	Bitter gourd	*Momordica charantia* L.	Rainy
Bhindi	Lady's finger	*Abelmoschus esculentus* Moench	Rainy
Kakadi	Cucumber	*Cucumis sativus*	Rainy
Muli	Radish	*Raphanus sativus* L.	All
Ogal	Buckwheat	*Fagopyrum esculentum* Moench	Summer
Sagiya Mirch	Capsicum	*Capsicum annum* L.	Summer and rainy
Bean	Bean	*Phaseolus vulgaris*	Summer and rainy
Tamatar	Tomato	*Lycopersicon esculentum* Mill.	All
Lai	Indian mustard	*Brassica Juncea* Czern & coss.	All
Methi	Fenugreek	*Trigonell foenum-graecum* L.	All
Palak	Spinach	*Spinacia oleracea* L.	Winter
Bakula	Field bean	*Vicia faba* L.	Summer
Matar	Pea	*Pisum sativum*	Winter
Aalu	Potato	*Solanum tuberosum* L.	Summer and rainy
Halang	Garden cress	*Lepidium sativum* L.	Winter
Baigan	Egg plant (brinjal)	*Solanum tuberosum* L.	Summer and rainy
Piyas	Onion	*Allium cepa* L.	Summer
Pinalu	Cocoyam	*Colocasia esculenta* L. (Schott.)	Winter
Gaderi	Taro	*Colocasia* Sp.	Winter
Turai	Ribbed gourd	*Luffa acutangula* L. (Roxb.)	Rainy
Chaulai	Garden amaranth	*Amaranthus tricolor* L.	Summer and rainy
Band Gobhi	Cabbage	*Brassica campestris*	Summer
Phool Gobhi	Cauliflower	*Brassica oleracea* Var.	Summer
Chichinda	Songe gourd	*Luffa cylidrica* (L.) M. J. Roem.	Rainy
Murch	Chilli	*Capsicum annum* L.	Rainy
Lasun	Garlic	*Allium sativum* L.	Summer
Dhaniya	Coriander	*Coriandrum sativum* L.	All
Haldi	Turmeric	*Curcuma longa* L.	Winter

Himalayan states, especially Himachal and Jammu and Kashmir attained tremendous achievement in this area. In the Alaknanda basin, there are various geographical locations which carry suitable geo-environmental conditions. These locations can be transformed into 'fruit belts'.

In the present case study area, 'the Alaknanda basin', cultivated land comprises 5.9 % of the total geographical area. The majority of the cultivable land is utilised for growing subsistence cereal crops, while the area under fruit production is considerably low. Of the total cultivated areas, fruits are grown only in 0.6 % of the total area and it is mainly in the mid-altitudes and highlands. When we compare economic viability of subsistence cereal crops and fruits, it is noticed that cultivating fruits is more viable than subsistence crops. It provides a base for enhancing livelihood, generating income and employment augmentation. The agro-ecological conditions and landscape of the basin have altitudinal diversity which in turn can ensure diversity in fruit production. Apple is grown in the highlands above 1,600 m. Citrus is cultivated between 800 and 1,600 m. Though, mango, guava, papaya and banana are grown in the low lying areas between 500 and 800 m, yet their proportion in production is negligible. Similarly, nut fruits, stone fruits, pears, peach, almond, strawberry and wild fruits are grown in the highlands, but the area coverage and production remains too low.

Distribution of Fruits

High diversity in fruit crops is observed throughout the whole basin. It varies vertically and horizontally. This study was based on the data collected from the entire region and drainage basins of the major rivers and their tributaries. The major rivers are the Alaknanda, Dhouli Ganga, Vishnu Ganga, Nandakini, Pindar and Mandakini. Here, types of fruit and their producing areas are given in the major agro-ecological zones. Table 4.3 shows the distribution of fruits according to altitude and river basin.

Table 4.3 shows the distribution of fruits (including wild fruits) in the Alaknanda basin, based on vertical and horizontal distribution. Agro-ecological zones vary from sub-tropical to cold (500–2,200 m). The major types of fruits are (a) apple-pear, peach, almond, apricot and nut fruits (1,600–2,200 m), (b) citrus-lemon, orange, mandarin and elephant citrus (1,000–1,600 m), (c) sub-tropical-guava, papaya and mango (500–1,000 m) and (d) wild fruits-pear, peach, kafal (*Myrica esculenta*), ber (*Ziziphus Jujuba*), hinsul (*Rubus ellipticus*) and bhamore (*Benthamidia capitata*) (500–2,200 m). Wild fruits play a vital role in subsistence economy and livelihood of the Himalayan people. A variety of wild fruits with enormous economic potential can be seen growing in the valley regions, mid-altitudes and the highlands of the basin. These fruits are used as food traditionally by the native people. However, their economic valuation has not occurred so far. The native people were able to bring value added products from some of these species during the last three decades and so. These fruits have also been used as a medicine while curing many diseases through traditional local health care systems.

Practices of Fruit Cultivation: Current Status

Practices of fruit cultivation have a long history in the basin. It has been in practice for centuries for domestic consumption. In the 1970s, the Government of Uttar Pradesh launched a programme for development of fruit cultivation in Uttarakhand (earlier known as U. P. Hills). A proportion of land—uncultivated and forest—was devoted and demarcated as ‘fruit belts’. The area is characterised by temperate climate. A large number of fruit plants were planted in the demarcated area. However, in many proposed fruit belts, this scheme was not successful. The reason behind this was the non-coordination between the state government and the state forest department. Many of the fruit belts were administratively and technically

Table 4.3 Distribution of fruits in the Alaknanda basin

Type	Altitude (m)	Agro-ecological zone	Producing area
Apple, pear, peach, almond, apricot and nut	1,600–2,200	Temperate cold	Dhouli Ganga, Vishnu Ganga, Upper reaches of Nandakini, Pindar and Mandakini Rivers
Citrus- lemon, orange, mandarin and elephant citrus	1,000–1,600	Sub-temperate to temperate	Mid-altitudinal regions of Nandakini, Pindar and Mandakini, higher reaches of lower Alaknanda basin
Guava, papaya and mango	500–1,000	Sub-tropical	Low-lying areas (river valleys) Nandakini, Pindar, Mandakini and Alaknanda
Wild fruits- pear, peach, kafal (*Myrica esculenta*), ber (*Ziziphus Jujuba*), hinsul (*Rubus ellipticus*) and bhamore (*Benthamidia capitata*)	500–2,200	Sub-tropical to temperate cold	In entire region of the Alaknanda River its tributaries

Source Adopted and modified (Sati 2004a, b, c, d)

under reserved forest area. The state forest department refused to convert the reserved forest area into fruit belts. Therefore, lands which were community land or uncultivated land belonging to the revenue department were declared and converted as fruit belts. Apple was the main item of these fruit belts, especially in the areas of Gwaldom-Lolti, Pothibasa-Duggal Bitta, Triyuginarain-Toshi and Kalimati-Janglechatti. This programme continued till the 1980s. Apple production from these fruit belts was high and it also penetrated the national and international market. The apple orchards are currently no more and they have been converted into agricultural land or land under cash crops. An owner of about 5 ha apple orchard, 'Shubhas Herbal Nursery (SHN)' Mr. Jodh Singh Badiyari was interviewed. The owner had extensive apple cultivation before the 1990s where about 50 quintals of apple were produced in a season. He informed that due to *disease of apple trees, the whole garden was destroyed.* Although, there is a horticulture training center at Gwaldom and the state government employed trainers to look after the orchards in the region, yet according to the owner of SHN, *the trainers never visited his orchard.* He requested trainers several times, but they did not turn-up. There are various reasons for failure of fruit cultivation. Climate change is one of them. The author, during his field visits, observed that the orchards of apple and citrus are abandoned. Dwarika Prasad Sati, a forester in the Forest Development Corporation said, *due to tremendous changes in climate, cultivation of fruits has completely vanished.* The areas where intensive cultivation of apple was carried out during the past are no more suitable or available for cultivation. This belt has been sifted greatly to the higher elevation. This is also the case with citrus and nut fruits and stone fruit belts. The extension workers of horticultural department and horticultural mobile teams could not do better for increasing production of fruits and help the poor farmers. Fruit orchard needs nourishment and high investment. It takes 4 to 5 years to grow. The farmers are marginal and poor. They are not able to manage their livelihood for 4 to 5 years without growing subsistence crops. They need financial assistance. In this context, government agencies can assist them financially.

Prospects of Fruit Cultivation

The whole basin is bestowed with suitable agro-ecological conditions, immense water resources and cheap human labour. These factors are opportunities and advantageous for cultivating fruits of various kinds. As the agro-ecological conditions vary from sub-tropical to alpine, fruits of sub-tropical, sub-temperate and temperate can be grown on a large scale with high quality and quantity. It is a common experience that the ecological conditions of the area are more suited to fruit cultivation rather than cereal farming (Atkinson 1889). Along with fruit cultivation, cultivation of tea and off-season vegetables can ensure a big leap to the regional economy (Sati and Kumar 2004). From the valley regions to the north border, sub-tropical humid and bio-climatic conditions change step by step into temperate, sub-temperate and alpine zones (Atkinson 1889) depending upon different factors, such as altitude, direction of slope and distance from snow-clad peaks. Of these factors, altitude is most important in determining the broad features of the climate of a particular region, such as temperature, moisture and rainfall. Fruit cultivation is even more overwhelming, owing to the presence of numerous mountains that transverse and tower above the surface and have relative relief ranging from 600 up to 7816 m. It is also varied from one altitude to another due to variation in the given factors.

Economy of the Himalayan region is based upon the cultivation of subsistence cereal crops. The production and productivity of these crops is considerably low. It does not meet even the daily food requirements of the people. Therefore, people of the region have migrated largely to the metropolitan city and to the plains. This has led to a severe situation of over and under-population. The whole region is ecologically fragile. Landslide and soil erosion is a common phenomenon. It reduces soil fertility and consequently low production of crops observed everywhere in the region. Cultivation of fruits on the hilly slopes of Garhwal region has dual impacts on the economy and landscape. It is a cash crop. Tremendous production of fruit crops will lead towards enhancing regional economy. Similarly, fruit trees conserve soil. The areas characterised by soil erosion should be demarcated and subsequently, plantation of fruit trees should be carried on.

Discussion and Conclusions

Practices of fruit cultivation are greatly influenced by government policies and participation of the people. The steps and attempts for growing fruits in the Himalayan region was unsuccessful due to two factors. Government decision to establish fruit belts in the temperate region was unsuccessful because of framing irrational planning and non-coordination with forest department. People's participation in the development process has a great role. There is history of failure of individual efforts towards diversifying and enhancing livelihood options. The marginal agricultural land is fragmented. There are various reasons for the failure of individual efforts; this includes fear of wildlife and threat from other owners who are less interested in fruit crop cultivation. This was observed by the author during his field visits. Under such circumstances, what are the steps to be taken for development of fruit cultivation? There is a general consensus among producers, researchers and academicians about community participation with modern innovation in farming systems. If an innovative idea is implemented by owners regarding initiation of enterprises-related fruit cultivation on community basis, sustainable development can be achieved. Like self-help groups, community participation in development activities should be ensured. Although there are various schemes launched for assisting farmers and organising community groups for development purposes, a major thrust is still required to assist the community groups as a whole and not at an individual basis. If a group of people participate together in cultivation of fruits, the entire scenario may be changed and development of the region may be ensured.

Potential of Dairy Farming

Dairy farming in the Alaknanda basin has vast potential. Suitability of environmental conditions for rearing lactating animals and suitable agro-ecological conditions for growing various types of grasses are largely found in the entire Himalayan region. During the past decades, lactating animals were reared and milk was produced for domestic consumption and also supplied to the local market. This trend of large-scale production of milk has considerably been reduced in due course of time because farmers in large numbers have given up this practice. Keeping suitability in the environmental conditions, the Government of Uttarakhand and the Department of Animal Husbandry initiated Intensive mini dairy projects (IMDP) initially for 5 years in each district of Uttarakhand.

Rearing livestock is a subsidiary occupation along with cereal farming in the Himalayan region. It consists mainly of rearing draught animals for manure and to plow agricultural fields. Domestic production of milk for self-consumption is also a characteristic feature of livestock rearing of this region. During the past decades, when the population pressure on the land was comparatively less and the annual production from subsistence cereal crops was able to meet the average grain requiremnet, rearing of draught animals was relevant because they were the main source of manure and draught power. Subsistence production of milk and milk-based products, mainly ghee, for household consumption and as well as its supply in the local market or within the villages was possible by rearing of lactating animals, i.e. cows and *murrah* buffaloes. Presently, the major species of lactating animals of this region are *Sahiwas*, *Red Sindhi*, *Jersey*, and *Halsteen Freejian* in cow and *Murrah* in buffalo. Each household on an average had a pair of bullocks, a cow and a buffalo. Along with meeting the need of draught power, plowing agricultural fields and providing manure households are able to produce milk for daily consumption. A number of households sell milk in the nearby market to earn their livelihood. Meanwhile, the trend of rearing draught and lactating animals has declined tremendously.

IMDP in Uttarakhand (2004–2005 to 2008–2009)

The Government of Uttarakhand had taken initiatives to start IMDP in all the 13 districts of the state initially for 5 years from 2004–2005 to 2008–2009. The main purpose of this scheme was to augment employment and to enhance economy of the rural people. For successful implementation of this project, scientific training was imparted to the beneficiaries to increase the milk production, income earning and augmentation of employment so that the migration of rural youth would be minimised. Table 4.4 shows target under IMDP in Uttarakhand for 2004–2005 to 2008–2009. This programme was largely successful in all the districts where it was launched.

Table 4.4 Target under IMDP in Uttarakhand (2004–2005 to 2008–2009)

	First year	Second year	Third year	Fourth year	Fifth year	Total
Establishment of mini dairy	1,350	1,450	1,500	1,550	1,600	7,450
Reserved for SC and ST	311	334	345	356	368	1,714
Employment generation						
Direct	4,590	4,930	5,100	5,270	5,440	25,330
Indirect	505	542	561	580	598	2,786

Source Dairy development, Uttarakhand, UCDFL, Haldwani (Nainital)

NABAARD, a statutory body, recommended establishment of a unit that would ensure that two animals are provided for each beneficiary. With the recommendation of this Unit, a loan of up to Rs. 30,000/- would be provided by the bank to the beneficiary. Subsidies in the form of transportation of animals (Rs. 2,500/-), initially for fodder (Rs. 560/-), insurance of animal and beneficiary (Rs. 2,520/-) and margin money (Rs. 3,000/-) would be given to each beneficiary for establishing a dairy unit. This department would provide the following services:

1. Technical investment facilities for beneficiaries having lactating animals.
2. Veterinary services at the village level.
3. Solving problems related to animal insurance.
4. Imparting scientific training to beneficiaries for animal management and fodder development.
5. Imparting training to beneficiaries for using organic farming and biogas plants and reducing chemical fertilizers in agricultural fields.
6. To ensure women participation for their empowerment.
7. To assist the schemes run by the NABAARD and cooperatives.
8. Benefiting landless labours, BPL families, and households totally dependent on agriculture.

Case Studies

CDPCUL Simli

This unit was started in November 1989 at Simli in Chamoli district. In the beginning, the entire Chamoli district was its administrative area for collection/ sale of milk and milk products; but now the unit has limited its activities mainly to six development blocks (Deval, Tharali, Narainbagar, Gairsain, Karnprayag and Pokhari). The daily collection of milk from these given areas is 720-l. The main objectives of this plant are to:

1. provide market at village level,
2. control milk price,
3. develop self-help groups,

4. provide loan facilities to the farmer for purchasing lactating animals,
5. empower women through giving them active participation,
6. arrange grain and green fodder for animals and
7. provide veterinary services to herders for rearing lactating animals.

Committees are set up at the village level, which include women and dairy herders and a chairperson that presides over all committees. The main function of the members of committees is to develop linkages with the cooperative society and the villagers. Furthermore, they also work for motivating the villagers to rear lactating animals and to become self-reliant. At Simli, seven employees are working in different capacities. A chairperson, three employees of the state government, and three women ad hoc workers constitute the CDPCUL. Recently, the output from the CDPCUL has decreased. As a result, the management has decided to terminate the services of three workers, including two women. It clearly demonstrates that the size of the society is reducing because of low output. Discussions with the personnel of the society identified that collection of milk from the different villages is not feasible because of lack of transportation facilities and less production. Meanwhile, the location of CDPCUL Simli is ideal for collecting and distributing milk and milk-based products.

BDF Bhararisen

A case study of BDF (Table 4.5) was carried out. It is located at the water divide of the Pindar and Ramganga rivers, about 30 km away from Karnprayag Township and about 5 km away from Diwalikhal on the way to Karnprayag-Ranikhet road. In the 1980s, the government of Uttar Pradesh established this farm at an altitude of 2,000 m. Ideal conditions such as climate, availability of fodder and gentle slope favoured this dairy farm. In 1990, per day milk production was approximately 3,000 l/day at the rate of 6 l/cow. About 250 employees were earning their livelihood from this farm. In 2008, the entire scenario changed. The area of dairy farm reduced from 1 to 0.8 sq km. Total number of cows reduced from 500 to 300. Most of them were calves. Milk production also marked a reduction to 1,500 l/day. The milk consumption area is limited to CDPCUL Simli. The author interviewed the employees of this dairy farm regarding this change; they responded that the government had withdrawn financial support and was intending to close the unit. Meanwhile, Bhararisen and its surrounding area has high potential in terms of suitable geo-environmental conditions, for rearing high-yield variety of animals. The farmers of this area are willing to continue this dairy farm at the cooperative level, if the state government extends financial-logistic and technical support.

Table 4.6 shows that the milk production and fodder requirement (per day) varies from one place to another according to the elevation. As the elevation increases the milk output also shows an increase. Buffalo milk production is higher than cow milk. Similarly, altitude and oak fodder play a vital role in the production of milk. The fodder requirement is largely satisfied in the high altitudes. Forests provide maximum fodder because there are large grass and forest lands surrounding the villages. The climatic conditions are feasible for production of milk

Table 4.5 Bhararisen dairy farm at a glance

No.	Characteristic features	1990	2008
1.	Location	2,000 m	2,000 m
2.	Area cover	1 sq. km	0.8 sq. km
3.	Total number of cows	500	300
4.	Total number of employees	250	100
5.	Per day milk production	3,000 l	1,500 l
6.	Milk consumption area	Between Karnprayag and Dwarihat	CDPCUL Simli

Source Primary collection

Table 4.6 Milk yields and fodder requirement (Per day)

Name of village (elevation) (m)	Accessibility from road (km)	Milk yield by cow (indigenous) (kg)	Milk yield by buffalo (indigenous)	Fodder used oak/ other (Wint./ Summ) (kg)	Crop residue (whole year) (kg)
Kimoli (1,900)	10	80	300	800/900	300
Prethi (1,600)	5	50	150	600/700	400
Ali (1,200)	0.5	20	25	Nil/200	400

Source Sati (2004a, b, c, d)

and rearing of different animals. At least the accessible areas, which are connected by roads, may be used for introducing hybrid animals. This process will boost the economy of the villages.

Prospects of Dairy Farming

Dairy farming is one of the economic activities upon which livestock rearing and farming communities in the mountain areas depend. It involves natural resources based-forest/rangelands, croplands, livestock breeds, feeding, health management, marketing and consumption of the product. Small-level dairy farming has enormous potential. It can contribute to family income, generate gainful employment especially for women, elevate living standards of the producer, fight malnutrition especially amongst children and enhance sustainable agriculture. Crop-livestock-forest integrity is a key factor in the sustainability of mountain livelihoods. Augmentation of dairy farming systems leads to enhanced performance of the overall production system.

Dairy farming is an integral part of the region's agriculture. Smallholders comprising most mountain farmers are accustomed to rear some animals as an essential component of the farming system. Among the various basic needs, the animals fulfill, milk is the most important as far as a family is concerned. Dairy animals are the best means to convert local vegetative biomass into useful products and work, and high value biodiversity into products of still higher value, such as dairy products fondly consumed by the masses (Singh 1996). That is why dairy animals occupy the predominant place in the herds of the region. Here, smallholders who make up the overwhelming majority have evolved two major systems of livestock management in the mixed crop livestock farming system: sedentary and migratory. Cattle and buffaloes are the only dairy species throughout the region. The use of goats and sheep as dairy animals is extremely rare. Grasses have great significance in fodder supplied to animals. *Napier, guni, dolani guchhi, rai* and *brom* grasses are suitable for fodder production here. These grasses are found between 500 and 2,500 m. Cultivation of these grasses will be more beneficial to herders of the region. The Uttarakhand Livestock Development Board (ULDB) is initiating to establish a 'Centre of Excellence for Fodder Grasses' in a selected 10 ha Van Panchayat land of 11 districts in the first phase. In the second phase, it will be extended to the other Van Panchayat areas of districts so that the problem of fodder can be reduced. ULDB has started to prepare compact field blocks of 5 and 2.5 kg to distribute to herders at a cheap rate.

Conclusions

If proper policies and programmes are formulated, dairy farming can be a subsidiary economic activity next to agriculture in this region. Similarly, it may enhance the livelihood of the populace. The agro-climatic conditions are feasible for rearing lactating animals. Furthermore, the extensive presence of pasturelands, temperate and alpine increases the possibility of intensive mini dairy farming. The major problems faced by farmers are the distance between the collection areas to the unit and lacking in number of refrigerators. Milk is collected from about 50–60 km distant areas in small-scale and often milk is spoilt. The dairy farming communities do not produce enough milk. Whatever they produce is used for self-consumption and seldom for selling in the local market. For promoting intensive mini-dairy farming the following suggestion are given:

1. Installation of refrigerators in each cluster of villages (10–12 in number) should be ensured to avoid wastage of milk.
2. Incentives should be given to milk-producing farmers for encouragement.
3. In each cluster of villages, a committee of milk farming community should be formed to handle all matters related to milk production. This would not only ensure the people's participation but would also be helpful for smooth functioning.
4. Involvement of local people in collection of milk and milk-based products should be ensured. This will augment employment to the poor rural youth.
5. Veterinary centres should be modernised and opened in each cluster of villages.

Prospects of Tea Cultivation

The Himalayan region has favourable agro-ecological conditions to cultivate tea mostly on the gentle slopes of mid-altitudes. With due recognition to these geo-climatic conditions, the British East India Company started tea cultivation during the 1830s. Initially, the tea cultivating areas were carefully developed and high grade quality of tea was produced; but later on, due to many impediments related to tea cultivation, production slowly showed a decrease till 1949. Currently, the Government of Uttarakhand has established a number of tea nurseries and gardens in the Kumaon and Garhwal Hills and tea production has again started.

The region is bestowed with suitable agro-ecological conditions, diverse forest resources, plenty of water resources, highly elevated snow-clad mountain peaks and charming landscape. Tea cultivation is carried out at Pauri and Chamoli districts. A case study of nine tea producing areas was carried out, which come under Nauti Sub-Projects (NSP) in Chamoli district. The other areas where tea is produced are Gwaldom in Chamoli district (oldest) Gadoli-Mandalkhal in Pauri district, and a long belt between Dehradun and Kalsi in Dehradun district. This study is mainly based upon interviews with extension workers, unskilled labours and local people engaged in this practice.

Tea Cultivation During British Regime

Commercial tea production in Uttarakhand was started during the British period. In 1824, Mr. Bishop Heber investigated the potential tea cultivating areas in the Kumaon region. This was followed by Mr. Ramley when he prepared a report on tea cultivation in the Kumaon and sent it to the East India Company in 1827. Consequently, Lord Venting set up a committee in 1834 to complete this mission. In 1835, around 2,000 plants of tea reached Uttarakhand from Kolkata. These tea plants were planted in Laxmeshwar in Almora and Bharatpur near Bhimtal. The cultivation was spread in the entire Uttarakhand gradually, and by 1837–1838, tea production started. It got recognition regarding its quality and the British Government appreciated this efforts. Until 1880 there were 63 tea gardens in 10,937 acres of land. About 500 workers were employed in tea gardens. By 1949, due to severe impediments related to tea cultivation such as export, transport, lack of markets, labour problems, lack of tea processing factories at local level, less popularity of tea in the producing regions and minimum interest of the owners towards tea cultivation the production got slowed down.

Tea Cultivation After 1990s

In March 1994, Public Investment Board (PIB) of Uttar Pradesh Government sanctioned Rs. 21.77 crore for Tea Development Projects in Chheedapani

Table 4.7 Proposed tea projects of Uttarakhand

Name of sub-project	Sanctioned target (ha)	Current situation (ha)
Kausani-Bhimtal sub-project	211	210
Nauti sub-project	200	114
Ghodakhal sub-project	200	12
Champawat sub-project	200	44
Total	811	380

Source Uttarakhand Tea Development Board (UTDB), Almora (2010)

(Champawat), Vijaypur-Kausani (Bageshwar), Bhimtal (Nainital), Gadoli-Mandakhal (Pauri) and Nauti (Chamoli) areas. PIB took a decision to establish tea gardens in 811 ha land with tea processing units in each sub-project. Till date, around 380 ha land is under tea plantation. The land is classified as Van Panchayat, Gram Panchayat and land of the farmers, which is taken on lease initially for 30 years. Table 4.7 shows details of the proposed tea projects of Uttarakhand.

Small-Scale Tea Development Programmes

Recognising the suitable agro-ecological conditions for cultivation of tea, the Government of Uttarakhand established a UTDB which is headquartered at Almora in 2004. As mentioned by the Board, the main objectives of establishing tea gardens in the state are to (i) harness the favourable agro-ecological conditions; (ii) enhance the economy of the people inhabiting in the surrounding areas of these tea gardens; (iii) ensure the people's involvement in tea cultivation to generate income and augment employment; and (iv) rehabilitation of degradable land and reduction of soil erosion through tea cultivation in these areas. The government collaborated with private sector to establish tea processing units in each area to facilitate the farmers to export their tea leaves. Land under *Van Panchayat*, community land and private land was possessed by the UTDB on lease initially for 7 years with a provision to return the land to the owner after 7 years. The motive behind was to prepare the land for cultivation of tea and when mass tea production started, the land would be handed over to the owners so that they could get direct and sustaining benefits. Training on tea cultivation was also imparted to the owners and their family members. In the initial 7 years, the family members could work in the tea gardens as daily labourers (per person per ha). Some other policies at the state level have been implemented to promote tea cultivation that comprise the National Sustainable Development Plan, Matching Grant Plan and Especial Component Plan. These schemes especially targeted the Schedule Casts and Schedule Tribes. Currently, the allotted area of 200 ha per garden has been reduced to 50–60 ha per garden with 10–15 km periphery (Sati et al. 2008).

After implementation of certain policies for the development of tea cultivation, the UTDB has achieved the following success; (1) plantation of tea in 380 ha land; (2) establishment of multiplication plats; (3) establishment of lab for soil testing in

Table 4.8 Tea Cultivation in Nauti Sub-Project

Name of tea garden	Land types	Year of establishment	Area in ha	No. of workers	No. of supervisors
Khageli	Van Panchayat	1997	10	24	2
Chulakot	Farm land	2003	2	2	
Chondali	Farm land	2000	7	13	1
Jakh-Lwenta-Pudiyani	Farm land	2004	5	7	1
Nauti	Van Panchayat/ Farm land	1995	16	32	4
Kanswan-Badet	Farm land	2002	4	6	1
Adi-Badri	Farm land	2000	11	16	1
Janglechatti	Farm land	1999	7	11	1
Kalimati	Van Panchayat	1999	2	3	1
Total			64	114	12

Source Sati and Mansoori (2009)

Bhawali (Nainital district); (4) establishment of nurseries; (5) imparted training for supervisors and labours (50 supervisors and 600 labours); (6) promotion for organic farming; (7) establishment of tea processing factories; (8) availability of global market for example as South Korea imported tea from Uttarakhand; (9) employment for 10 permanent and 56 ad hoc employees, 810 households benefited and 52 % employment has been given to women; and (10) a research centre has been established at Kausani in 2003 by the GBP University of Agriculture and Technology, Pantnagar for the promotion of tea cultivation.

A case study of tea cultivating areas of Nauti sub-project in Chamoli District was carried out, which includes Khageli, Chulakot, Chondali, Jakh-Lwenta-Pudiyani, Nauti, Kanswan-Badet, Adi Badri, Janglechatti and Kalimati with its processing centre located at Bhatoli; land types, year of establishment, area in ha, number of worker and number of supervisor. Table 4.8 shows the detail description of tea cultivation in Nauti sub-division.

Total area of tea gardens in Nouti sub-division was 64 ha. Total numbers of workers were 114 and supervisors were 12. Apart from this, three field assistants were appointed. Of them, one is a permanent employee (on deputation) and the other two are on ad hoc basis. The supervisors were appointed on ad hoc basis with monthly salary ranging from Rs. 2,800/- to 4,000/-. Field assistants, who also work as ad hoc employees, receive Rs. 6,000/- per month while permanent field assistants receive Rs. 12,000/- per month. Workers engaged in collecting tea leaves receive Rs. 79 per day. Bhatoli is a service centre where a tea nursery was established. A private tea processing unit has also been established here. This unit buys the green leaves and processes for sale. An ISI level was given by the Government of Uttarakhand to this tea processing unit. Tea is sold locally, regionally and globally. Tea production has obtained global markets due to its high quality.

Workers, supervisors and field assistants were interviewed regarding potential of tea cultivation and income earned from it. Rupees 64,800/- is paid as salary to supervisors and field assistants per month. Further, workers (114 Nos.) total salary payment per month is Rs. 2,70,180. It is Rs. 3,34,980/months plus cost of maintenance of tea gardens, Rs. 40,19,760 annually. The output from the tea cultivation was estimated. The off-season production of green tea leaves was 20 kg/ha/day, which was 1280 kg/month. During the tea growing season (4 months of monsoon), production increased up to 100 kg/ha/day which was 6,400 kg/month (25600 kg during monsoon period). Annual production therefore reached to 35,840 kg. Per kg cost of green tea leaves was Rs. 25. Therefore, the annual output from tea cultivation was Rs. 8,96,000/-. Cost–benefit assessment shows that at present the output from tea cultivation is tremendously low but in due course of time, the output will be multifold as the employees observed. They informed that productivity would reach its maximum in the near future after the remaining saplings attain production capacity. It is estimated that if the production increases up to 100 kg per ha land, the annual production from 64 ha tea garden would be 76,800 kg and annual income would be Rs. 19,20,000/-. If the area under tea gardens is increased in the near future, the prospects of tea cultivation will reach tremendously high levels. The farmers involved in tea cultivation directly and indirectly are optimistic about the prospects of tea cultivation, but they are unhappy about the statutory fixed, low level of wages.

Tea cultivation is a promising sector and can play a significant role in the sustainable development of the region. In the areas where agro-ecological conditions are suitable, tea cultivation may be promoted with a wider perspective. This will enable sustainable livelihoods of rural poor. Gentle slope and moderate climate are favourable conditions for tea cultivation. These conditions are abundantly found in the mid-altitudes of the Himalayan region.

Policies need to be reframed and implemented for livelihood enhancement and economic development. Master trainers, extension workers and other governmental agencies should be active and they should monitor tea cultivation from time to time. Without government support, it is difficult to achieve the target. Similarly, active community participation will lead to proper development of tea plantations.

Worldwide research in the field of livelihood enhancement shows that diversification in agriculture farming may increase the livelihoods and income of the rural poor. In mountain regions where options for enhancing and diversifying livelihood are less, the cultivation of tea may be a promising sector subjected to availability of favourable landscape and suitable agro-ecological conditions. The present study region has suitable agro-ecological conditions and best landscape for tea cultivation. This study reveals that at present output from tea cultivation is tremendously low but its future prospect is considerably high. For sustainable tea cultivation in the study area, suitable locations should be identified and local participation should be ensured.

Non-timber Forest Products

Non-timber forest products (NTFP) can be promoted in degraded land with the willingness of the Village Forest Committees (VFCs) through a participatory approach or the concept of User Group within the Village Forest Development Committee (VFDC) for gradual adoption of the short, as well as long-term model. Initial ideas can be adopted from central and state government sponsored forestry and forest-based programmes. The Ministry of Environment and Forests, Government of India is promoting conservation and development of NTFP, including Medicinal Plants under National Afforestation Programme (NAP). The 100 % Central Assistance Scheme and Projects sanctioned are executed over a 5-year period through Forest Development Agencies (FDAs), registered as societies under the State Societies Registration Act.

NTFP is defined as forest products of biological origin other than wood as well as services from forests and allied land uses. FAO defines "Non-Wood forest products consist of all goods of biological origin other than wood in all its form, as well as, services derived from forests or any land under similar use". Shiva defines: "All products obtained from plants of forest origin and host plant species yielding products in association with insects and animals, including their parts and, items of mineral origin except timber be classified as NTFP. The main NTFP are medicinal plants, wild fruits, honey and honey bees, organic materials, bamboo, oak bark, natural dyers, ringal (a variety of bamboo) and environmental services. Plantation of ringal[1] on a large scale should be ensured. The Oil and Natural Gas Commission (ONGC) has recently initiated a scheme for plantation of ringal in Chamoli and Rudraprayag districts (Fig. 4.1).

Resin Industry

The resin industry has potential to enhance the regional economy. The raw materials and other favourable conditions are largely available in the Himalayan region. Resin locally known as '*Leesa*' is extracted from the pine trees. It is collected in big containers and supplied to the depots. From the depots, it is supplied to places where resin industries are located. It is processed for making different important items.

---

[1] A small variety of bamboo widely used for making households goods mainly for the collection of crops and their products.

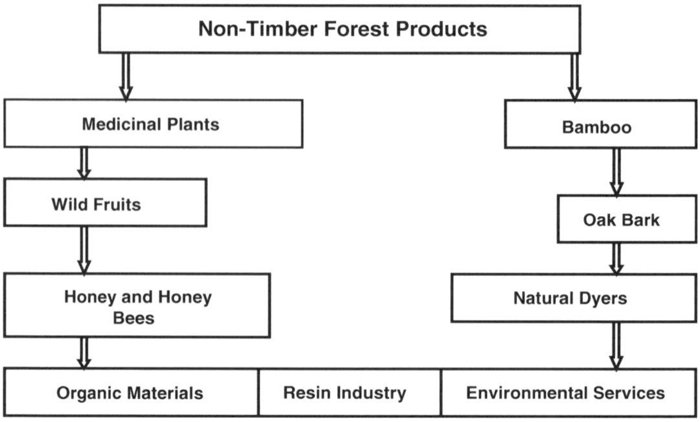

Fig. 4.1 Non-timber forest products

Definition of Resin

A semisolid substance secreted from the sap of some plants and trees. It is used in varnishes, paints, adhesives, inks and medicines.

Hard resin: a hard translucent resin ranging in color from amber to dark brown and derived from the sap, stumps or other parts of pine trees. It is used in the manufacture of varnishes and other products and to increase friction, for example, between the bow and strings of stringed instruments.

Oil from resin: a thick yellowish sticky liquid distilled from resin and used in making varnishes, inks and other products.

Distribution of Pine Trees

In the case study area, the Alaknanda Basin forest covers about 67 % of the geographical area. Of this, 40 % is covered by pine forest, densely found in the major sub-tributaries of the Alaknanda River. Dense pine forest can be seen in the Pindar, Nandakini, Mandakini, Alaknanda rivers and their sub-tributaries between 800 and 1,600 m altitude. The extensive and dense pine forests here provide a suitable base for the resin industry.

Case Study of Lastar Gad Sub-watershed

A case study of Lastar Gad sub-watershed (a major tributary of the Mandakini River) was carried out. About 50 % land is covered by pine forest. The detail description of resin extraction in Lastar Gad sub-watershed is as follows:

Table 4.9 Details of Resin Industry in the Lastar Gad

Name of area	North and South Jakholi Range
Collection point	Saral Van Parisar
Season of getting resin	March to November
Production per day	17 kg
Production per season	20 thousand tin
Number of segments (blocks)	24
Number of contractor	24 (one contractor in each segment)
Number of workers	5–7 worker in each segment
Non-skilled labour	Local and Nepali immigrants
Contractor	Local
Number of pine trees in each block	5,000 (estimated)
Area of each block	200 ha
Location of resin industry	Earlier at Tilwada now in Rishikesh
Products from resin	Tarpin oil, Biroja, Koltar Paint, Oil for furniture and glass

Source Primary collection

Table 4.9 shows general features and details related to resin extraction in the Lastar Gad sub watershed. The area where resin is extracted from pine trees is in the North and South Jakholi Range. Extracted resin is collected in Saral Van Parisar, which is located about 5 km downslope of Jakholi town. Resin is extracted between March and November each year. Per day production of resin is 17 kg with 20,000 tin per season (a tin contains 15 kg resin). The whole sub-watershed is divided into 24 segments (blocks) by the Forest Department. A contractor and 5–7 workers are engaged in each segment. The workers engaged in extraction of resin are uneducated local people and Nepali immigrants. Contractors are locally appointed by the Forest Department. The number of pine trees in each segment is about 5,000 while the area under pine trees is about 200 ha (each segment). The resin industry was earlier set up at Tilwada in Rudraprayag district now shifted to Rishikesh. While discussing with the employees of the Forest Department, contractors and labours engaged in resin extraction activities, a common suggestion arose that the resin industry should be shifted to Tilwada instead of Rishikesh. Elaborating on this issue, the respondent stated that establishing a resin industry in Tilwada will pave the way to enhancement of the regional economy and livelihoods of the local people. It will augment direct employment to the local people who should be the real beneficiaries from this industry. Involvement of the local people from resin extraction to finished products will surely manifest a way towards livelihood enhancement. A similar strategy should be adopted in other areas such as the Pindar River Basin, the Nandakini River Basin, the Mandakini River Basin and the areas located at the mid-altitudes of the Alaknanda River basin (ARB). These are areas where very dense pine forest is found. Resin industry should be established at Narain Bagar for the middle Pidar basin, Deval for the upper Pindar basin, Ghat for the Nandakini Basin, and Rudraprayag for the Mandakini and Alaknanda Basins.

An effort was made to get cost–benefit analysis from resin extraction. A container having 15 kg weight costs from Rs. 1,000 to 3,000 depending upon distance between the areas of resin extraction to the place where it is gathered. It is sold at Rs. 3,000–4,000 per container to the factory owner. The benefit goes to the Forest Department. It is obvious that if the entire work from extracting of resin to the final products is handed over to the community people, benefits would go directly to them and they would be able to enhance their livelihood.

Wild Fruits

Wild fruits play a vital role in the subsistence economy and livelihood of the Himalayan people. A variety of wild fruits with enormous economic potential are found in the valleys, the mid-altitudes and the highlands. These fruits are used as food traditionally by the natives. However, their economic valuation has not occurred so far. The native people were able to bring value-added products from some of these species during the past decades, but currently, this practice has tremendously reduced. These fruits have also been used as medicines for curing many diseases (Table 4.10).

Medicinal Plants and Herbs

India is endowed with a rich biodiversity in plants and is recognised as one of the world's top 12-mega diversity hotspots. It possesses rich flora that include about 45,000 species and many are accredited with medicinal value. Over 15,000 species are used in different systems of health care in Asia (7,000 in China and 8,000 in India). However, available information shows that 1,700 species are used in Classical Indian Systems of Medicines. Ayurveda uses 1,200, Siddha 900, Unani 700, Amchi 600 and Tibetan 450 medicinal plants for their preparations. A majority of the raw materials for these preparations are collected from forests, whereas some medicinal plants are cultivated. Despite the close relationship between forests and pharmaceuticals, little effort has been made to maintain, manage and develop technology for conservation of these medicinal plant resources from Indian forests. The estimated 95 % of medicinal plants, collected in India, are from the wild and the process of collection is destructive because of the unscientific means of collections employed. An estimate of the parts used by Ayurvedic industries are roots—29.6 %, leaves—25.8 %, bark—13.5 %, wood—2.8 %, whole plant—26.3 %, and rhizome—4 % and rest are seeds, flowers, etc. A major part of the high range Himalayan plants are wild harvested and many of these are close to extinction due to over-harvesting or unskilled harvesting, e.g. nardostachys jatamansi and aconitum species.

Table 4.10 Potential wild edibles in the Alaknanda Basin

Local name	Botanical name	Location (height in m)	Harvesting period	Potential use
Bel	Aegle marmelos	Up to 1,200	July–August	Ripe fruit is eaten fresh and used for making juice and squash. Sacred in Hindu religion. Medicinal use: astringent, digestive and stomachic
Guriyal	Bauhinia purpurea	Up to 1,300	Feb–March	Flower buds are cooked as vegetable and also pickled. Used in dysentery, piles, diarrhea, and ulcer. Leaves are used as fodder
Semal	Bombax cieba	Up to 1200	Feb–March	Flower buds are used for making pickle and cooked as vegetable. Leaves are used as fodder
Kingore	Berberis asiatica	600–2,700	June–July	The berries are edible, sweet/sour in taste and used for making juice and squash. Roots form a reputed drug in Ayurvedic medicine
Bhamore	Benthamidia capitata	1,500–2,500	Nov–Dec	Fruits are edible and used for making jam
Lingra	Diplazium polypodies	1000–2000	June–Nov	Fronds of the plant are used for making pickle and also cooked as vegetable. It is useful for patients suffering from cough, asthma, fever, stomachache, dysenteric, dyspepsia and diarrhoea
Gweain	Eleagnus latiifolia	1200–3000	Sept–Oct	Fruits are edible and used for making juice and squash. Good for cough and bronchitis. Ability to fix atmospheric nitrogen
Anwala	Emblica officinalis	500–1,500	Dec–Feb	Fruits are edible and used for making pickle, murabba, squash and juice. Medicinal and nutritious used for curing of cough, anemia, peptic ulcer, diabetic, asthma and bacillary dysentery. Rich source of vitamin 'C'
Timla	Ficus auriculata	1,000–1,200	July–August	Fruits are eaten, making pickle and cooked as vegetable
Gular	Ficus glomerata	900–1,500	Feb	Ripen fruits are eaten; unripe fruits are used for making pickle and cooked into vegetable. Leaves are used for fodder
Bedu	Ficus palmate	500–1,800	May–June	Fruits are edible, used for making pickle and cooked into vegetables. Medicinally used for digestive disorder. Leaves are used for fodder
Khaina	Ficus semicordata	900–1,400	Oct–Nov	Fruits are used for making pickle. Leaves are used for fodder
Ames	Hippophae rhamnoides	2,000–3,000	Nov–Dec	Food, medicinal and cosmetic uses. Curing cold and cough
Sahtut	Morus serrata	1,200–2,600	July–Sept	Leaves used for rearing silkworms. Fruits are used for making juice and squash

(continued)

Table 4.10 (continued)

Local name	Botanical name	Location (height in m)	Harvesting period	Potential use
Kaphal	Myrica esculenta	1,400–2,000	May–June	Fruits are eaten and used for making juice and squash
Ghingaru	Pyracantha crenulata	800–1,800	June–July	Fruits are edible and used for making jam, sauce, juice and squash. Cures cough and cold
Burans	Rododendren arboretum	1,200–2,400	March–April	Flowers are used for preparation of juice and squash
Hinsul	Rubus ellipticus	1,000–2,000	May–June	Fruits are edible and used for making jam
Amara	Spondias pinnata	700–1,500	Dec–Feb	Ripen fruits are eaten and processed into juice, sauce and squash. Curing dysentery and diarrhoea
Bhatmoliya	Viburnum mullaha	1,500–2,800	Nov–Dec	Fruits are edible and used for making juice and squash
Ber	Ziziphus Jujuba	Up to 1,500	Nov–Feb	Edible and used for making juice, murabba and squash

Source Compiled by author

The state of Uttarakhand has been declared as a *'Herbal State'* by the government in 2003. It is rich in terms of biodiversity of medicinal plants with thousands of species; however only about 320 species are recognised in terms of their medicinal value. The Forest Department claims to have knowledge of about 175 species, which are being commercially extracted and traded. But the district-wise inventory is yet to be completed. Experts, however, estimate that in terms of value, the state is well positioned to generate revenue of about Rs. 1,000 crores per annum through Herbs and Medicinal Plants (H and MP) in the raw form. Most of the species (about 95 %) are found wild in the forests. The forest management has traditionally been timber oriented and tree centric. Hence, the H and MP sector has been deprived of the similar attention of shrubs and herbs vis-à-vis trees. However, due to ease of collection, the tree species of medicinal plants got earlier attention, compared to their counterparts, i.e. herbs and shrubs. Traditionally, the native people and *Bhotia* tribe of Uttarakhand were developing H and MP for their livelihood. This resource not only provides those primary herbal medicines but also provides nutrition to cattle population in particular and also contributes substantially to their income. Even today, the majority of the population derive their income from H and MP. Besides, there has been a significant shift globally towards H and MP for medicines and Medicare materials with a concretion. This is considered as more patient friendly and eco-friendly. Declaration as *'Herbal State'* has necessitated to define the operational, functional mechanisms formulating action plan for sustainable development of an H and MP Sector in the State. Recently, cultivation of herbs mainly *kutki* and *kut* has been largely done in Ghaise village of Tharali taluk. This has been carried out with the help of HAPPRC, Srinagar Garhwal. About 25 HHs have been benefited with an amount of Rs. 87,000 during the last 2 years. Initially 32 HHs were involved with this practice but 7 HHs withdrew because of payment-related issues. With the interference of HAPPRC, the payment was made.

Most of the H and MP grow in the wild as natural components of vegetation. But this traditional base is shrinking due to informal supply chain, overexploitation, population pressure, unsustainable practices and biodegradation. There are restrictions on extractions and procurement from wild; yet, the trade is active secretively and results in overexploitation of the resources and this finally leads to unsustainable practices. In the absence of a scientific system of collection and fostering regeneration of such plants, several species have been completely lost, have become endangered or are on the verge of extinction to varying degrees. This is serious genetic erosion and leads to the loss of biodiversity of the region. Research in H and MP has focused on their biologically active compounds and evaluating their remedial properties, but little attention has been paid to the sustainable management of these plants in the State. The Uttarakhand Government is yet to take effective measures to provide sustainable management of marketing system and appropriate policies for conservation/protection of H and MP. Notwithstanding the measures of propagation and cultivation of H and MP, initiated extraction rather than collection from the wild sources through unofficial trading systems is still going on unabatedly on a large scale. The magnitude of illegal trade

Table 4.11 List of sacred H and MP species of Garhwal Himalayas

Scientific name	Vernacular name	Beliefs/Uses
Emblica officinalis	Amla	Sacred tree
Aegle marmelos	Bail	Sacred plant
Musa paradisica	Banana	Used in rituals
Ficus benghalensis	Bargad	A sacred tree
Cedrus deodara	Deodar	Sacred tree
Artemisia sps.	Dhoop or Kunju	Used in rituals
Cynodon dactylon	Doob	Used in rituals
Mangifera indica	Mango	Used in rituals
Azadirachta indica	Neem	Sacred tree
Quercus spp	Oak	Sacred tree
Prunus cerasoides	Paiya	Used in rituals
Ficus religiosa	Peepal	A sacred tree
Pinus roxburghii	Pine	Used in rituals
Xanthoxylum achanothopodum	Timroo	Sacred tree
Ocimum sanctum	Tulsi	A sacred herb

Source Adopted from Anthwal et al. (2006) and Sati (2013)

is high, the majority of gatherers remain unidentified and their actions are guided purely by profit motives. This creates great threat to the conservation efforts (Table 4.11).

Initiatives for Sustainable Development and Management of H and MP

After declaration of Uttarakhand as a *Herbal State*, the government has taken initiatives for sustainable management of H and MP in a phased manner. The government has appointed the Agriculture and Processed food products Development Authority (APEDA) as the nodal agency to promote setting up of Agri Export Zones (AEZ) in two phases. Under the first phase, six districts—Chamoli, Dehradun, Haridwar, Pithoragarh, Udhamsingh Nagar and Uttarkashi, are being covered. In this phase, emphasis is placed on ten high value species and these plants would be cultivated on about 500 ha. In the second phase, the area under cultivation would be increased and additional districts would be brought under the aegis of AEZ and the variety of medicinal plants would also be diversified. This is being done with support from the Infrastructure Development Finance Company Ltd. to boost the exports and enhance India's share in the world market. The state has established a Herbal Research and Development Institute (HDRI) at Gopeshwar in district Chamoli as nodal agency to monitor developmental issues and for inventory of H and MP species in the state.

Industries within the state are encouraged to establish Research and Development Centres as 'Centres of Excellence'. A Committee would work out the modalities and norms. Independent certification agencies could monitor the indicators and certify the system as sustainable. For example, institutions like the GBPIHED/GBPUA and

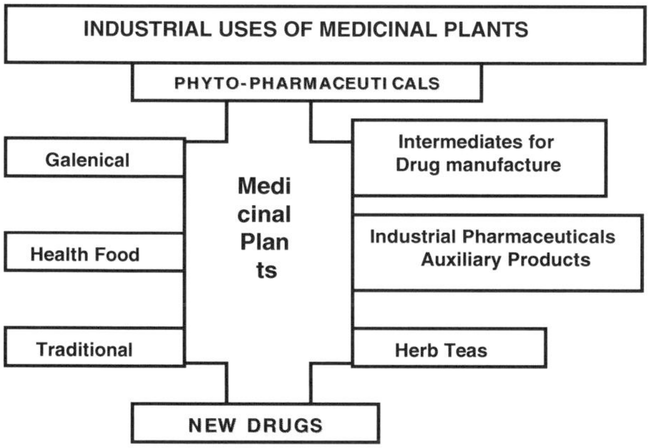

Fig. 4.2 Industrial use of medicinal plants *Source* FAO Bull. No. 11, 1997

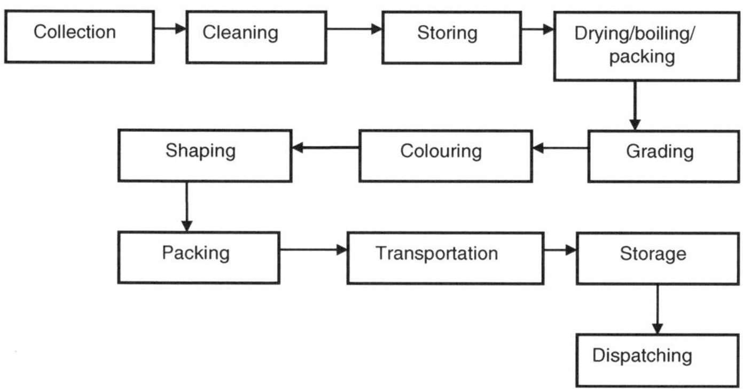

Fig. 4.3 Minimum post-harvest steps required for marketing of H and MPs

T/HNBGU/ICFRE within the state could act as an impartial agency for such work. They could further extend training in certification issues with the help of WWF/ World Bank Alliance initiative. A number of NGOs (Including NRIF) and other autonomous institutions in the state or outside could be identified to undertake certification. The development of group certification schemes will help to lower the cost of certification (Extracted and synthesized from NRIF 2004) (Figs. 4.2, 4.3).

Bees and Beekeeping

Beekeeping contributes enormously to income generation through bee products and pollination services. It has been estimated that through pollination services to agriculture and surrounding flora, beekeeping helps communities to generate income equipment to 14 times the investment required (Free 1993). Beekeeping, especially at commercial levels may be an option for enhancing livelihood. It may substantially contribute to the livelihood for small and marginal farmers. In the upland regions of the ARB, where a feasible ecological condition exists throughout the year, apiculture can be promoted extensively. The upland has also the privilege of having flowering season throughout the year. Already, in many areas of the upland people are involved in apiculture and are producing considerable quantities of produce.

Beekeeping in the traditional way is a centuries-old practice in the Himalayan region. Bees make their artificial habitat mostly on the roof of buildings, which is locally nomenclatured as '*jala*'. Honey has multiple uses. It is a supplementary food, has medicinal purposes and helps in pollination. It is also has an eco-friendly practice and is consumed domestically. As the economic viability of beekeeping is very high, and the ecological conditions for its sustainable practices are quite suitable, the region may lead in its production. It is noticed that there has been a considerable decline in this practice during the last decades. The government has initiated several steps to set up stations for beekeeping both in the Kumaon and Garhwal Himalayas.

'Jalikot' in Nainital district is the headquarter involved in training, coordination and supervision. Talwari in Chamoli district works parallel to assist the farmers of Chamoli district for beekeeping. The other centres are at Pithoragarh, Almora, Nainital, Pauri, Kotdwar, Uttarkashi and Chiniyalisaud. In Dehradun, five centres, viz. Rishikesh, Balawala, Nathuwala, Nakraunda and Kunwawala provide technical services.

Three wooden boxes with ten frames are available in each centre. Five personnel from the horticultural department are deputed in these centres for producing honey. They include the development investigator, supervisor, sub-supervisor, field man and peon. These boxes are made of wood of various colours and stories for the safety of frames. Half kilogram of sugar and 1 l water is weekly required for each box during the winter and rainy seasons. During flowering seasons, i.e. autumn and spring, bees collect their food from the flowers at daytime. 10 kg honey is produced from one frame in a year. The main species of bees is *Italian seeds of malifera*. The production of bees from this species is 15–20 kg/frame/year. The reproduction period is flowering seasons, when about 200 to 300 eggs/day reproduced. Within a year when the frame is overlapped the group is divided into two under a separate woman leadership. Development workers are required at this crucial time for their rehabilitation.

The major problems with the production of bees in this area are (i) the species are indigenous and consequently, production is low; (ii) market is improper and inaccessible, therefore the bee producers do not get even the total cost of input; (iii) although there are training centres set up in each district and trainers are appointed, yet producers do not get proper training for improving honey production; and (iv) traditionally practiced beekeeping is risky. Despite these problems, beekeeping has many prospects such as (i) honey is used as medicine; (ii) it is widely used as food supplement; (iii) if market is available, it can increase the income level of the people involved, and (iv) it does not need much investment, hence even marginal and small farmers can practice this.

Conservation and Promotion of Indigenous Honeybees

Notwithstanding the achievements, there is still much to be done in terms of institutionalizing and replicating the successes of the programme across the region. Capacity building will continue to be an integral part of any future programme. In addition, increasing honey production to meet the growing demand for organic honey will reduce poverty. To counteract the reduction in the efficiency of pollination services requires new thinking and approaches to efficiently utilise and develop available honey bees' resources. The impact of climate change is another important issue that requires attention. Indigenous approach includes farmer-managed selection and multiplication of *Apis cerena*, which leads to an increase in honey productivity from a mere 2 kg per hive per year to an average of 6 kg at project sites and a maximum of 13 kg at one site (Ahmad et al. 2008). The adoption of better management practices by farmers resulted in a reduction in absconding, greater resistance to disease and parasites, and an increase in colony numbers and colony strength. Capacity building efforts by the programme were focused on partnership development, networking, curriculum development and training.

Natural Dyes

In the Himalayan region, natural dyes provided by plants and animals have been in use for imparting different shades, generally to wool and cotton, from time immemorial. The various ethnic communities, including Bhotiyas of the Garhwal Himalaya residing in the border regions, are well known for their traditional expertise in making a range of woollen garments. Before 1962, there was trans-border trade between India and former Tibet and the import of wool was the major source of income for the Bhotiya's woollens-based, indigenous cottage industry. But after 1962, trade was stopped due to conflicts between China and India. However, due to the visit of a sizeable number of pilgrims and tourists to the Badrinath and Kedarnath shrines, the Bhotiya community still has a market to sell woollen products, so that this age-old tradition of preparing woollen products and processes to make them colourful through indigenous methods are still exisiting in the remote areas.

Table 4.12 Plants and their parts used in preparing natural dye

Plant species	Part used	Colour produced
Khukhuyinya, Rumex nepalensis	Root	Yellow
Dolu, Rheum australe	Root	Yellow
Akhrot, Juglans regia	Fruit cover	Camel
Kapasi, Corylus jacquamontii	Fruit cover	Camel
Kingod, Berberis spp.	Root	Yellow
Archa, Rheum moorcroftianum	Root	Yellow
Bajar Bhang	Root	Brown
Geranium nepalense	Root	Red

Source Kala (2002)

Indigenous Approach of Wool Dyeing

The process of wool dyeing starts once the threading of wool is over. Normally, raw wool has two colours, black and white; only the white coloured wool is used for dyeing purpose to give different shades. The woollen material is dusted and ashed thoroughly before dyeing. The fruits of *Sapindus mukorosii* (Reetha) are used traditionally for washing the woollen threads. The roots or fruit covers of collected or stored plant species are washed thoroughly with tap water and then sun-dried. The dried plant material is powdered and mixed with water to prepare a solution. The solution is heated in a vessel until it boils. The wool threads are then slowly immersed in the solution for dyeing. For dyeing 1 kg of wool, around 4 l of water is boiled with approximately 50–60 g of fresh root. The woolen threads are stirred thoroughly for a long time to ensure thorough and uniform soaking. Sometimes, a small quantity of ash is also poured in the prepared solution for better colouring. Once the dyeing is over, the woolen threads are taken out from the boiling solution for drying. During drying, direct sunlight is avoided for retaining brightness of the colours (Table 4.12).

The Bhotiya tribal economy and their indigenous practices, evolved to maintain this industry, were based on personnel interactions and keen observation of their environment. However, use of natural dyes started to decline gradually after the discovery of synthetic dyes in 1856. In view of the wide availability of colours through synthetic dyes, the traditional natural dyes which offered only a limited range suffered a severe setback. Besides, the decline in the woollen cottage industry is also due to lack of vision and ignorance of the traditional heritage and values. Difficulties in getting raw material (desired plants) and preparation of natural dyes are among the many causes for the decline in use of these indigenous techniques. However, extraction of natural dyes is cost beneficial and natural dyes are user-friendly products. It can also lead to employment generation at the village level. If serious efforts are made, natural dyes can prove to be a boon to the textile industry, which can market its products to the European countries and to the USA. The abundant forest biomass can be used for manufacturing dyes on small-scale bases, especially in the villages. Unfortunately, this indigenous practice of using natural dyes has almost vanished even from the remote villages of the Uttarakhand Himalaya (Extracted and sysnthesised from Kala 2002).

References

Anthwal, A. et al. (2006). Sacred groves: Traditional way of conserving plant diversity in garhwal Himalaya, Uttarakhand.

Atkinson, E. T. (1889). *The Himalayan Gazetteer* (Vol. 2, p. 369). Delhi: Cosmo Publication.

Free, J. B. (1993). *Insect pollination of crops*. London: Academic Press.

Indian Council of Agricultural Research. (2002). Agricultural Research Data Book, ICAR, 2004.

Kala, C. P. (2002). Indigenous knowledge of Bhotiya tribal community on wool dyeing and its present status in the Garhwal Himalaya, India. *Current Science*, *83*(7), 814–817.

Natural Resources India Foundation (NRIF). (2004). Mechanism for sustainable development & promotion of herbal and medicinal plants in the State of Uttaranchal. SER Division, Planning Commission. Government of India.

Ramakrishnan, P. S., Saxena, K. G., Swift, M. J., & Seward, P. D. (Eds.). (1993). *Tropical soil biology and fertility research: South Asian context*. Dehra Dun: Oriental Enterprises.

Sati, V. P. (1993). Cropping pattern in the hill environment of the Garhwal Himalaya. (A case study of a village). *National Geographer, 28*(1), 31–37. www.mtnforum.org

Sati, V. P. (2004a). Systems of agricultural farming in the Uttaranchal Himalaya, India. *Journal of Mountain Science, 2*(1), 76–85. http://www.imde.ac.cn/journal

Sati, V. P. (2004b). Systems of vertical horticulture in the Alaknanda Basin of Garhwal Himalaya. *Annals of NAGI, 24*(2), 49–58.

Sati, V. P. (2004c). *Horticultural development in hills: A case for the Alaknanda Basin.* New Delhi: Mittal Publications, (Ph. D. Thesis). www.vedamsbooks.com/38930

Sati, V. P. (2004d). *Uttaranchal: Dilemma of plenties and scarsities.* New Delhi: Mittal Publications. www.vedamsbooks.com/38930

Sati, V. P. (2005a). Vertical zonation of horticultural farming in the Alaknanda Basin of Garhwal Himalaya, India. *Journal of Mountain Science, 2*(4), 319–328. http://www.imde.ac.cn/journal

Sati, V. P. (2005b). Natural resource conditions and economic development in the Uttaranchal Himalaya, India. *Journal of Mountain Science, 2*(4), 336–350. http://www.imde.ac.cn/journal

Sati, V. P. (2013). Cultivation of medicinal plants and its contribution to livelihood enhancement in the central Himalayan Region. *Advancement of Medicinal Plant Research, 1*(2), 17–23.

Sati, V. P. & Kumar K. (2004). *Uttaranchal: Dilemma of plenties and scarsities.* New Delhi: Mittal Publications. www.vedamsbooks.com/38930

Sati, V. P. & Rawat, M. S. S. (1993). Problems of agriculture ecology and their management in the Pinder Basin. *National Geographer, 28*(2), 143–148. www.mtnforum.org

Sati, V. P., Maithani, D. D. & Kumar, K. (2008). Prospects of tea cultivation in Uttarakhand. *Green Farming: A Journal of Agricultural Sciences*, *1*(9), 41–43. www.greenfarming.in

Sati, N. & Mansoori, I. K. (2009). *Pilgrimage tourism.* New Delhi: Mittal Publications.

Singh, V. (1996). *Draught animal power in mountain agriculture: A study of perspectives and issues in the central Himalayas, India.* Discussion Paper. Series No. MFS 98/1 (p. 74). Kathmandu: ICIMOD.

Singh, B. (1998). Vegetable production under protected conditions: Problems and prospects. *Indian Society of Vegetable Science Souvenir: Silver Jubilee, National Symposium* December 12–14 (p. 90). Varanasi, U.P. India.

Singh, N., Diwedi, S. K. & Paljor, E. (1999). *Ladakh Mein Sabjion Kei Sanrakshit Kheti.* Regional Research Laboratory of DRDO, Leh. Published By D.R.D.O., Leh. Published By D.R.D.O. 56 A.P.O.

Chapter 5
Livelihood Analysis

Livelihood comprises the capabilities, assets (including both material and social resources) and activities required for a means of living. It is sustainable when it can cope with and recover from stresses or shocks and maintain or enhance its capabilities and assets both now and in the future, while not undermining the natural resource base (DfID 1999; Scoones 1998; Carney 1998). The author has attempted to elaborate the changes in the lifestyles of the rural communities from the past to the present. The prime focus is on traditional livelihood and its current pattern. As observed in many cases, the livelihood pattern in mountain regions is in transition. To understand the traditional pattern and modern situation of livelihood in the study region, primary data were gathered and analysed. Subsequently, a study was carried out on the current livelihood patterns of the communities in the clusters of villages of Chamoli and Pauri districts through a detailed field survey. Meanwhile, the focus of the field survey was to get detailed data from the primary sources, knowing about the earning pattern of rural communities from different livelihood sources and the expenditure for meeting their daily requirement.

Traditional Livelihood Pattern

The traditional lifestyle of the people living in the Himalayan region is almost similar to other highland communities all over the world. It has four very distinctive characteristics: (a) a combination of activities including agriculture, pastoralism and seasonal processing of forest produce to sustain the family economy; (b) seasonal transhumance involving pastoralism, with several combinations—using the favourable conditions provided by the seasons at different altitudes to optimise the resource availability for the family economy. Here, pastoralism is characterised by a nomadic lifestyle with summer and monsoon spent in the '*bugyals*' (alpine grasslands), and winter in the '*Bhabar*' region at the foot of the outer mountain range. Families engaged in agriculture split up and the more able-bodied spend the summer and monsoon in seasonal shelters called '*chhappers*' (Sati 2008a, b, c, d, e, f) used for large ruminants like cattle and buffaloes, farming

V. P. Sati, *Towards Sustainable Livelihoods and Ecosystems in Mountain Regions*, Environmental Science and Engineering, DOI: 10.1007/978-3-319-03533-8_5,

fertile land on the forest margins referred to as '*danda*' with crops like potatoes and amaranth and more recently herbs. Both are located near a ridgeline high above the village. During this period, the rest of the family remain in the village and farm the 'sera'—broad, irrigated terraces in the valleys for paddy, and the '*ukhar*' or '*upraon*'—unirrigated marginal lands around the village for coarse grains like millets and pulses; (c) complete dependence on forests (oak and temperate grasslands) for all elements of life, e.g. firewood, fodder, fibre, medicines, supplementary foods, water and soil conservation. Leaf fodder for cattle and leaf litter for cattle bedding provide the inputs for agriculture. Both are further processed to make air-dried compost, the only fertiliser used for subsistence agriculture, which is the major factor behind stable yield of traditional crops.

The rural economy traditionally consists of a combination of activities, in which agriculture and animal husbandry provide the base of the subsistence economy. Utilisation of non-timber forest products (NTFPs) provide the base for the market economy, with marketable products in demand locally. In recent years, the major trend emerging has been diversification of livelihoods beyond the traditional lifestyle of farming, pastoralism, collection and processing of NTFPs. This diversification clearly shows a steady rise in the importance of off-farm incomes in contributing to the family economy as compared to farm incomes, and a corresponding rise in the household expenditure on food. Interviewees identified this diversification as being the key to the continued economic viability of the village and family economy. It is clear that the rural population is realising that just glorifying traditions can no longer sustain a growing population because the carrying capacity of '*jal, jungle, and jameen*' (water, forest and cropland resources) to sustain the present population has exceeded.

Livelihood Analysis

Family income, expenditure and the proportion of families dependent on different livelihoods were analysed as case studies from 12 villages of Khanda Gad sub-watershed in Pauri District and 17 villages from Kewer Gadhera sub-watershed in Chamoli District. The objective was to identify the major trends in livelihoods. The analysis was based on interviews with farming families, artisans, families belonging to the Scheduled Castes and women were also interviewed. Data were gathered about levels of annual family income from various sources, family expenditure on household needs/other categories and the interviewees' perceptions of the proportion of the population dependent on various livelihoods in the village. General observations about the overall standard of living in the villages were also cross-checked with interviewees and incorporated into the findings. Livelihood analysis of case studied cluster of villages was carried out. A formula for livelihood sustainability index was prepared by the author to know the sustainability in livelihood of the people.

$$\begin{aligned}\text{Livelihood Sustainability Index (LSI)} &= X\left(x^1 + x^2 + x^3 + x^4\right) \\ &\quad - Y\left(y^1 + y^2 + y^3 + y^4 + y^5\right) \\ &= XY\end{aligned}$$

where X stands for household income that includes x^1 = income from farming, x^2 = income from business, x^3 = income from government service and x^4 = income from labour. Further, Y stands for household expenditure, y^1 = for food, y^2 = for shelter, y^3 = for cloth, y^4 = for education and y^5 = for others. $X - Y$ is equal to XY, i.e. saving. The objective towards formulation of this index was to understand the level of sustainability in livelihood. Based on this, LSI at village level and household level was prepared. It was inferred from the study that the trend of saving money is high even among low-income group people.

Household Income (*X*)

The major sources of livelihoods comprise agriculture, daily wage labour, jobs, artisanry, pensions, animal husbandry and NTFP collection. Levels of total annual family income were calculated by taking the weighted average of the annual family income of the respondents. Similarly, the proportion of income from each livelihood category was calculated by the same method. However, agriculture is primarily subsistence in nature, and in subsequent analyses the income from agriculture represents only the cash income and not the subsistence value. Similarly, for other categories of livelihoods, only the cash incomes are considered.

The households in each village were placed into four categories according to their work type and income; the income of each village seemed similar with minor variations. This is because, as Table 5.1 shows, the income level in each village depends on the earning from remittances (money orders). The households working only in agricultural fields have low level of income and this group constitutes a considerable portion of the society. There is a group that comprises people who have migrated to the metropolitan cities/the plains of Ganges and have left their agricultural land abandoned. Table 5.1 shows the village-wise proportion of the total income contributed by the various sources of livelihoods. Incomes from farming, business, government services and labour were found to be the most important constituents of household income. Agriculture constituted one of the major sources of household income. In many villages, mostly located in the highlands, the major source of income was from farming of subsistence crops. Another major source of income is from remittances. People of the region have migrated in large numbers to other parts of India in search of better livelihoods. They work in different institutions to manage their livelihoods sustainably. A few people have business ventures. Uneducated people seasonally work in agricultural fields as landless labour. They work in small service centres nearby the villages. However, their proportion is considerably less.

Table 5.1 Average income of the villages of Khanda Gad and Kewer Gadhera sub-watershed

Village	X^*				
Khanda Gad sub-watershed	x^1	x^2	x^3	x^4	X
Margaon	14,000	Nil	46,000	2,000	62,000
Shrikot Khanda	12,000	15,000	50,000	1,000	76,000
Margadna	10,000	Nil	52,000	1,000	63,000
Bhitai Malla	18,000	Nil	32,000	3,000	53,000
Kaldung	13,000	Nil	42,000	2,000	57,000
Dhanak	22,000	Nil	28,000	4,000	54,000
Gaduwa Gad	11,000	10,000	44,000	2,000	57,000
Kamand	25,000	Nil	25,000	5,000	55,000
Dov	24,000	Nil	26,000	5,000	45,000
Shiyar Malla	20,000	Nil	30,000	4,000	54,000
Rawat Gaon	12,000	Nil	48,000	1,000	61,000
Bhitai Talla	14,000	Nil	40,000	1,000	55,000
Kewer Gadhera sub-watershed					
Kewer Talla	10,000	20,000	52,000	1,000	83,000
Kewer Malla	15,000	2,000	40,000	2,000	59,000
Bhagoti	8,000	25,000	60,000	–	93,000
Ratni	6,000	Nil	38,000	1,000	45,000
Keshwan	6,000	Nil	42,000	–	48,000
Gadseer	12,000	Nil	40,000	2,000	54,000
Bunga	13,000	Nil	32,000	3,000	48,000
Jhijodi	10,000	Nil	30,000	4,000	44,000
Ali	4,000	3,000	46,000	–	53,000
Leguna	8,000	Nil	40,000	1,000	49,000
Bedula	8,000	Nil	40,000	1,000	49,000
Chirona	6,000	Nil	50,000	–	56,000
Kaub	15,000	8,000	60,000	2,000	85,000
Naini	6,000	8,000	58,000	–	72,000
Swan Malla	15,000	Nil	28,000	4,000	47,000
Swan Talla	12,000	Nil	34,000	3,000	49,000
Kimoli	16,000	Nil	32,000	5,000	53,000

Source Primary collection, X^* value is in rupees

Household Expenditure (*Y*)

If we look closely at the expenditure patterns in the study villages, it can be observed that in almost all the villages, nearly a quarter of the income is spent on the purchase of food grains. Household supplies in the form of sugar, spices, edible oil, salt and cooking gas rank second in household expenditure. The figures shown in Table 5.2 are reflective of a trend that there is an increasing dependence upon the outside markets for supply of food grains and that the villages are no longer

Table 5.2 Village wise average household expenditure

Village	Y^*					
Khanda Gad sub-watershed	y^1	y^2	y^3	y^4	y^5	Y
Margaon	18,000	12,000	16,000	14,000	Nil	60,000
Shrikot Khanda	20,000	13,000	20,000	16,000	2,000	71,000
Margadna	15,000	10,000	15,000	14,000	2,000	56,000
Bhitai Malla	12,000	10,000	12,000	11,000	1,000	46,000
Kaldung	14,000	15,000	12,000	10,000	Nil	51,000
Dhanak	12,000	10,000	11,000	13,000	1,000	47,000
Gaduwa Gad	12,000	10,000	12,000	11,000	1,000	46,000
Kamand	13,000	12,000	10,000	10,000	Nil	45,000
Dov	11,000	7,000	6,000	10,000	Nil	34,000
Shiyar Malla	13,000	12,000	10,000	10,000	Nil	45,000
Rawat Gaon	15,000	10,000	15,000	14,000	2,000	56,000
Bhitai Talla	10,000	11,000	9,000	9,000	Nil	49,000
Kewer Gadhera sub-watershed						
Kewer Talla	20,000	10,000	14,000	19,000	2,000	69,000
Kewer Malla	10,000	11,000	9,000	9,000	Nil	49,000
Bhagoti	20,000	15,000	16,000	27,000	3,000	81,000
Ratni	11,000	7,000	6,000	10,000	Nil	34,000
Keshwan	11,000	7,000	6,000	10,000	Nil	34,000
Gadseer	13,000	12,000	10,000	10,000	Nil	45,000
Bunga	11,000	7,000	8,000	12,000	Nil	38,000
Jhijodi	11,000	7,000	6,000	10,000	Nil	34,000
Ali	12,000	10,000	12,000	11,000	1,000	46,000
Leguna	12,000	10,000	10,000	11,000	Nil	43,000
Bedula	12,000	10,000	10,000	11,000	Nil	43,000
Chirona	13,000	12,000	10,000	10,000	Nil	45,000
Kaub	20,000	13,000	20,000	16,000	2,000	71,000
Naini	20,000	10,000	14,000	19,000	2,000	69,000
Swan Malla	11,000	7,000	6,000	10,000	Nil	34,000
Swan Talla	11,000	7,000	8,000	12,000	Nil	38,000
Kimoli	10,000	11,000	9,000	9,000	Nil	49,000

Source Primary collection, Y^* value is in rupees

self-sufficient in terms of food grains. This may have wide-ranging implications in terms of food security. This also reflects a general trend that there is growing emphasis on the cultivation of cash crops and hence a decline in the availability of food grains for self-consumption. This trend is also symbolic of the increasing consumerist nature of the villages, wherein people no longer consume the traditional food grains.

Table 5.2 presents the village-wise average expenditure. The major portion of the income is spent on food, shelter, clothes and education. The rate of expenditure varies from village to village. In villages located in the low-lying areas, expenditure on education is considerably higher than the villages located in the highlands. For other commodities such as clothes, shelter and food, the expenditure is more or less similar.

Emerging Trends

The economic analysis of the 29 study villages strongly suggests that an increase in the rural household incomes would not necessarily lead to an increase in the rural household savings. In fact, with an increase in incomes, the expenditure levels also seem to increase and as a result the monetary savings remain more or less static. More exposure with the outside world has inculcated the same consumerist trends as prevalent in the cities and towns of the plains. As a result, the quality of life of the village communities may not improve significantly with the increase in rural incomes. Attributable to jobs, people tend to spend a greater proportion of their income on providing quality education for their young ones. This is a clear indication that people are not greatly satisfied with their traditional agrarian economy and rural lifestyles. Given an opportunity, they want themselves as well as their children to secure good jobs and perhaps even leave their homelands so that they could enjoy urban lifestyles. However, in the case of the more remote villages such as Jhijoni and Kimoli people are more content with what they have; there is lesser desire to go outside and hence less importance is placed on education. As traditional lifestyles have been gradually replaced by consumerist values, a pre-occupation with jobs is observed. Faith in traditional spirituality has eroded tremendously with the growing influence of the cash economy. The steady increase in roads connecting remote areas has certainly brought convenience and ease of access. At the same time, they have transformed to modern, globalisation-influenced values, eroded local culture and many positive traditions.

Village-Wise Comparison of Income and Expenditure

We can conclude that income from agricultural, its allied practices and from remittances are not enough to ensure a satisfactory, sustainable livelihood. The rural educated youth, who do not get employment according to their choice, prefer to remain unemployed. They do not want to go back to work in the agricultural fields and rearing animals. This is a chronic situation in the villages. The education sector is renowned as an industry because of its potential to provide teaching jobs. In the valley regions, the unemployed youth are able to work in the quaternary sector as they started their businesses in the small service centres mostly located on the road heads. The ratio of income to expenditure is equal; if the household income is high, then the expenditure is also high and vice versa. In both the cases the rate of savings is considerably low. A small amount of savings is invested towards ceremonies. But now, a small portion of savings is also used for education.

References

Carney, D. (Ed.) (1998). *Sustainable rural livelihoods: What contribution can we make?* Department for International Development.

DFID (2000). *Achieving sustainability poverty elimination and environment, Strategies for achieving the international development targets*, report from DFID. Retrieved from: http://www.dfid.gov.uk/Documents/publications/tspenvironment.pdf

Sati, V. P. (2008a). Traditional intramontane mobility in Garhwal Himalaya: A survey of subsistence practices in the Pindar Basin, Uttaranchal. *Singapore Journal of Tropical Asia*, *29*(2), 173–185. A peer reviewed journal of Blackwell Publications. http://www.blackwell/journal/sjtg_v29_i2/sjtg_333

Sati, V. P. (2008b). Farming systems and strategies for sustainable livelihood in eritrea. *African Journal of Food Agriculture Nutrition and Development*, *8*(2), 219–237. A peer reviewed quarterly journal. http://www.ajfand.net/issue19/PDFs/Sati2505.pdf

Sati, V. P. (2008c). Natural resource management and food security in the Alaknanda Basin of Garhwal Himalaya, India. *Peer reviewed proceedings of 'Tropentag 2008' an International Conference*, Germany.

Sati, V. P. (2008d). Natural resource management and food security in the Garhwal Himalaya. *ENVIS on Himalayan Ecology*, *16*(2), 6–16. http://gbpihed.nic.in/envis/HTML/vol16(2)/vpsati.htm

Sati, V. P. (2008e). People's participation in forest resource management in the Uttaranchal Himalaya. *MF Bulletin* Vol VIII Issue 2 July 2008. http://www.mtnforum.org

Sati, V. P. (2008f). *Natural resources management and sustainable development in Pindar Valley Himalayas*. Dehradun: Bishen Singh Mahendrapal Singh.

Scoones I. (1998). *Sustainable rural livelihoods: A framework for analysis*. Working Paper 72, Institute for Development Studies, Brighton, UK.

Chapter 6
Development of Eco-Tourism and Hydroelectricity

Potential of tourism is enormous in the whole Himalayan region. Tourism has become a significant part of the local economy and major source of income. In tourism also, this region has diversity in nature. Plenty of natural scenic beauty spots, places for adventure and pilgrimage present a unique combination; it provides the opportunity for extensive and diverse tourism. It has opened new avenues for the local people to rise above the poverty levels. Creating new infrastructural facilities especially hotels/home stay etc., generate employment opportunities for the local people at largescale. This is the positive aspect of tourism which gave new opportunities and avenues to generate money. The other aspect is the deterioration of the environment. Degradation of forestland in and around the tourist places, garbage problems due to booming up of teastalls and shops and mass tourism are the major environmental problems. In 1990s a wide range of debates on development of tourism took place because during this period the awareness towards environment got a momentum. Consequently, a new concept of ecotourism emerged.

Development of hydroelectricity has also tremendous potential to enhance livelihoods of people of this region. The Himalayan region is endowed with abundance of water resources. The Ganga, the Sindh and the Brahmaputra systems are among the biggest river systems of the world. These rivers and their tributaries provide livelihoods to half of the population of the south Asian countries. They are glacial fed, and are perennial in nature. In this chapter potential of the development of tourism and small-scale hydroelectricity projects have been discussed. A case study of the Alaknanda Basin was carried out to assess the potentials of tourism and hydroelectricity.

Tourism Types and Development

Tourism is a smog free industry and one amongst biggest source of revenue generation worldwide. The economic development as establishment of the new avenues, employment generation and earnings of foreign currency are the major

V. P. Sati, *Towards Sustainable Livelihoods and Ecosystems in Mountain Regions*,
Environmental Science and Engineering, DOI: 10.1007/978-3-319-03533-8_6,

features of tourism. The scenic and panoramic view of the area further accelerated the scope of tourism. Tourism in a broader sense has existed for a long time in the Alaknanda Basin in the form of pilgrimages, located in the highlands and they are sacred places for the Hindu sanctity. Mountaineering, trekking, river rafting, rock-climbing and skiing are part of the adventure tourism in this region. These forms of mass tourism have not only influenced the socio-economy of the region but also influenced the environment and the local social structure. Recently, adventure tourism has been developed in Auli (district Chamoli) as skiing and river rafting in the lower Alaknanda Basin. These forms of tourism have been developed as international winter sports. Besides this, natural tourism, cultural and pilgrimage tourism also exist.

This region has potential of all the above-mentioned three types of tourism. Tourism practices range from visiting the highlands sacral pilgrimages for the peace, penance and salvation; unique cultural attractions, hiking, skiing, river-rafting, trekking and rockclimbing to see the panoramic view of the natural locales. During the past years, the modern (western) trend sports have gained momentum in the Himalayan region in the forms of rafting, kayaking, canyoning, rockclimbing, mountain-biking, bungee-jumping and paragliding (NTB 2000). Details of tourism types in the Alaknanda Basin are as follows:

Pilgrimage Tourism

Pilgrimage tourism is a centuries old practice in this region. The worshipping of the four *Dhams*, holy rivers and nature deities has its roots in the Aryan culture and was later integrated into Hinduism. The whole Himalayan region has an important spiritual meaning for Hindus as a 'sacral space' (Grotzbach 1994). Badrinath, Kedarnath, Yamnotri and Gangotri are the world famous pilgrimages, have their own religious importance and are wellknown for their natural and scenic beauties as well. It is believed that the Pandavas, the great rulers of India during the ancient period, passed through this region and built the famous Kedarnath temple, while on the way to the heaven. Similarly, there is a story of the Adi Guru Shankaracharya of Kalidi, who built the Badrinath temple to reunite the disintegrated Hindu followers. As the above description shows, these pilgrimages are the symbol of Hindu sanctity that kept the unity among the followers of Hinduism. Pilgrims, not only from the Indian sub-continent but also from the world, visit these *Dhams* every year. As the data show (Table 6.1) the number of pilgrims and tourists arrived, during the two different years—2000 and 2010 in these pilgrimages and tourist places have shown tremendous increase. Highest increase in number of visitors was noticed in Nainital with 84.7%, during the two periods. It was followed by Yamunotri (69.2 %), Gangotri (36 %) and Haridwar (35.5 %). Besides, the famous Sikh pilgrimages, the Govind Ghat (1829 m) and Hemkund Sahib (4329 m) are located in Chamoli district, on the way to Badrinath.

Table 6.1 Tourists flow in the major pilgrimages and natural locales

Places	Altitude (in meter)	Year		
		2000[a]	2010[b]	Increase in %
Haridwar	314	5324639	8261078	35.5
Nainital	1939	39696	258713	84.7
Mussoorie	2000	850746	980921	13.3
Gangotri	3293	208407	329111	36.7
Badrinath	3415	695332	921950	24.6
Kedarnath	3553	300000	400014	25.0
Yamunotri	4421	88672	287688	69.2

Source[a] Sati (2004a, b, c, d), [b] Different sources

Thousands of Sikh pilgrims visit these places every year. A survey was conducted by Nirmala and Mansoori (2009), showed that within 14 h duration, 14,980 pilgrims visited Kedarnath out of which, about 45 % people used pony or mule or wooden palanquin the rest by walk in a peak month of June, 2008. Meanwhile to Badrinath pilgrimage, about 22,000 pilgrims visited in a day, in the same month. Bathrinath pilgrimage is approached by the road transportation therefore; the pilgrims are outnumbered than to Kedarnath pilgrimage, where a pilgrim has to trek about 14 km. During April 2005–March 2006, total number of visitors in Uttarakhand was 11,619,457, out of which, 9,13,0397 were domestic, 1,09,084 were foreign and 2,379,994 were day visitors (Nielsen, AC ORG-MRG 2007).

Natural Tourism

Natural tourism developed after the establishment of hill towns. In the late 1830s, the hill towns became more attractive for the civilian residents of India, especially for the colonial middle and upper class. This was due to the fact that the hill stations provided the opportunity to escape the hot pre-monsoon months and the summer monsoon and stay in a more pleasant region with a beautiful landscape (Grotzbach 1994). Mussoorie (founded in 1827), Nainital (1839), Almora, Pithoragarh and Ranikhet (1889) became the major centres of attraction. These hill towns have magnificent landscapes, attract tourists every year at a large scale and contribute tremendously to the economic development of the region. Modern adventurer tourism started in the 1960s. This led to increase in the number of tourists in these towns. Walter (2000) stated that adventurer tourism in the Himalayan region started after Sir Edmund Hillary and Tenzing Norgay climbed the Mt. Everest and made the region popular in other parts of the world.

Adventurer Tourism

Adventurer tourism got impressive position as winter sports skiing has been developed at Auli, river rafting in the Alaknanda River, trekking in the various natural locales and pilgrimages such as Tungnath (3680 m), Rudranath (3553 m), Madhamaheshwar (3286 m), Triyuginarayan (1980 m), Roopkund (5029 m), Gomukh (4000 m), Purnagiri (3000 m), highland lakes, national parks, wild life sanctuaries and mountaineering in many of the snow-clad peaks.

Wildlife Tourism

This region is famous for the national parks, wildlife sanctuaries, bird sanctuaries and biosphere reserves. Forest covers about 65 % of the total geographical area. Among the famous national parks of India; Rajaji, Govind Ghat, Valley of Flowers and Corbett National Parks are located in this region. Rajaji National Park, located in Dehradun and Haridwar districts, is world famous for conserving elephants, while Corbett National Park (in Nainital District) is known for conserving tigers as the tiger project of India was started from here. Nanda Devi Biosphere Reserve (in Chamoli District) is declared as a world heritage site for the presence of the famous *Kasturi mirg* (an endangered wild dear) and also for the huge floral and faunal diversity. It is one of the biospheres in the world, which is located in the highest elevation. A famous bird sanctuary—'Asan Barrage' is located in Dehradun District, where the Siberian cranes take shelter during their winter migration. This wetland also provides shelter for numbers of permanent and seasonal migratory birds. Although, the national parks, wildlife and bird sanctuaries can be developed for the eco-tourism at a great level and it has already been developed to a certain extend yet, its pace is considerably slow. It needs a holistic approach for the sustainable development of park and wildlife tourism that will enable the region to achieve a high economic growth. This may also be possible through clubbing the package of pilgrimage tourism with park and wildlife tourism, since these locations are adjacent to each other. Tour operators can play an important role in this regard.

Cultural Tourism

Cultural tourism can play a significant role in the socio-economic development of the Himalayan region as it has rich culture and customs. It is believed to be an abode of Gods and Goddess thus, it is known as *'Dev Bhumi'*. Here, diversity in culture can be seen in every small geographical unit and in all walks of life. Therefore, it can be divided into many cultural realms. The two distinct cultures—

Garhwali and Kumaoni, divide the Uttarakhand Himalaya into two principal cultural realms. Further each cultural realm has numerous sub-cultural units. The language, food habits and clothing are remarkable different from one cultural realm to other. The fairs and festivals are celebrated in all months of the year. They coexist with the nature and seasons. Here, animals and trees are worshiped from time immemorial. This tradition of worshiping animals and trees manifests the way to conserve the nature. The uniqueness in culture and customs can promote cultural tourism greatly. Maha Kumbh in Haridwar and Nanda Devi Rajjat are among the prominent festivals, celebrated here.

Tourist Places of Religious Importance

Panch Prayags

There are numbers of places where the major rivers meet with the Alaknanda River. These places are known as 'Prayags'. It is interesting to note that these 'Prayags' are only found in the Alaknanda River. The first Prayag starts with the confluences of Saraswati and Vishnu Ganga (also known as Alaknanda River) at Mana village, 3 km away from the famous Hindu pilgrimage 'Badrinath'. This is known as 'Keshavprayag'. This Prayag is not included in the famous 'Panch Prayag' of Garhwal. According to the religious wisdom of Hinduism, the first Prayag starts with 'Vishnuprayag' where 'Dauli Ganga' confluences with the Alaknanda River near Joshimath town. Traveling from Joshimath to Karnprayag, Nandprayag town is located where Nandakini River confluences with the Alaknanda River. Pindar River meets with the Alaknanda River at Karnprayag. There is a myth that Karna, son of Kunti did penance here about 3200 B.C. A Karna Kunda is located at the meeting point of the two rivers on a huge stone where the pilgrims offer tributes to late Danveer Karna. Rudraprayag is located at the confluence of the Mandakini and the Alaknanda rivers. Although, there is another 'Prayag', i.e. Sonprayag where Son Ganga meets with the Mandakini River, yet very few description is available in the Hindu religious wisdom about this Prayag. The last and famous Prayag among Panch Prayags is 'Devprayag', where the Bhagirathi and Alaknanda rivers meet. This place has a religious importance because the Bhagirathi and Alaknanda rivers are called 'the Ganga' from here.

Panch Kedars

At 3553 m, Kedarnath temple is located in Okhimath tahsil of Rudraprayag district. The Mandakini River originates from Kedar peak and flows from the right side of Kedarnath temple. It is not only the place of religious importance but it has also a picturesque landscape. For around 6 months, the Kedar temple and its surrounding areas remain covered with snow while snow fall occurs even during the summer, which is the main season for pilgrims. There is a small township which is summer camp for the people of the low-lying areas. Few people out-migrate here with their animals mostly during rainy season, when they find extensive pastureland for gazing their animals. For around 6 months, the Pandas (religious Gurus) and the owners of small hotels or shops move here to manage their small-business establishments. According to a myth, the temple was established by the Pandavas before departing to *Swarga*. This myth is supported by the idols of five Pandavas, Draupadi and Kunti, which are stalled inside the temple. The Adi Guru Shankaracharya was travelling to visit the temple of Kedarnath but while on the way, he died at Bhairav Jaap. In the Garbhagriha (pivotal area), a flat triangle Ligna of Lord Shiva lies where the people make *pritkarma* at the time of Darshan. Panch Kedar includes Kedarnath, Tungnath, Madhyamaheshwar, Rudranath and Kalpeshwar. Kedarnath pilgrimage is one of the most important pilgrimages where about 14 km trekking is required by the pilgrims from Gauri Kund. It is a tough trekkingpath. The peak season is May and June, and September and October every year.

Panch Badris

Panch Badris include Badrinath, Bhavishya Badri, Yogdhyan Badri, Adi Badri and Adinath/Bridha Badri. Lying on the lap of Nar and Narayan Parvats (mountains), Badrinath is located on the right bank of Vishnu Ganga (Alaknanda River) at 3100 m. It is one of the four pilgrimages of India. The others are Puri in Orissa, Rameshwaram in Tamilnadu and Dwarka in Gujarat. Adi Gugu Shankaracharya established this temple. It is a seat of Lord Vishnu. Mainly, the Hindu pilgrims visit Badrinath Temple. Badrinath temple is opened in the second week of May and closed in the mid-October. Remaining the year, it is covered in snow. High number of pilgrims visit Badrinath temple every year.

The infrastructural facilities such as transportation, lodging and boarding are considerably less in these two pilgrimages. Though, road transportation is available up to Badrinath temple, the condition of road is not good. From Helang to Badrinath, the condition of road is bad but it is worst between Joshimath and Badrinath, where one way traffic system prevails because of the narrow road. The capacity of lodging and boarding is low, the expenses in these pilgrimages are therefore high and the pilgrims of low income group cannot afford it. Income from pilgrimage tourism is comparatively less than from the places of natural tourism in this region, while the pilgrims are outnumbered.

Tourism has multiple impacts on local economy, as it increases employment opportunities and living standard of the local people consequently increasing the national income (Sati 2004a, b, c, d). With the number of visitors increasing dramatically, the total amount of money spent by them also increases. In fact, it is a major source of income as Rangan (2000) reported that Garhwal earned between 10 and15 % of its revenue from taxes and levies imposed on pilgrims each year. In India, tourism is the second-largest source of foreign currency behind the gem and jewelry business (TED n.a.). The money spent by the tourists has diverse effects on the local economy. It stimulates the economy and induces the so-called multiplier-effect viz. jobs are created, capital is accumulated and local business ventures increases. Those businesses, in turn, employ people as guides or workers, which thereby benefit indirectly the local economy. But a part of money can also be used to improve the local living standards through better health care, education and building structure. The economic impact of tourism can be noticed along the roadsides from Rishikesh to Badrinath and Kedarnath, where mushrooming of small teastalls, dhabas, can be observed. A study (Sati 2013) reveals that the villages that lie on the major routes to the pilgrimages are now fully dependent on the income from the tourists because they provide them accommodation and other basic facilities to facilitate their stay.

While tourism in the study area is an important source of livelihood for the people and revenue, the environmental implications can not be neglected. Land degradation is the most common effect of tourism, particularly in hill resorts. The demand for fuel wood the primary fuel for restaurants and dhabas, is much higher during tourist season, especially in the higher locations of the basin, leading to further pressures on the forests (UEPPCB 2004). In addition expansion of tourist areas causes degradation of adjoining green belts, as most of the tourist centres are located in places with good natural vegetation.

Hydroelectricity Development

The Himalaya is bestowed with rich biodiversity and provides wide varieties of natural resources including the life sustaining water to the Indian subcontinent. In the last few years significant increase in the number of proposed hydroelectric projects has been witnessed in various river basins. The enormous hydroelectric potential of Uttarakhand has made it synonymous with *Urga-Anchal* (Power State). During the 1980s, there were 22 proposals of hydroelectric projects and many of them have commenced. These macro-hydroelectric projects are mostly commencing in the areas where human settlements and productive agricultural patches are densely found. There are totally 220 power projects proposed in Uttarakhand. Out of which, 52 are of large size, 36 medium and 132 small-scale projects. From each macro project, 100 MW power productions are targeted. Similarly, from medium project, 25–100 MW productions are proposed and 25 MW productions are proposed from small-scale projects. There are seven dams, which have more than 200 m height. The state has about 30,000 MW hydroelectric generation capacities. At present, 8,000 capacity powerhouses are under construction. The major projects where generation of power has already started are Dhauliganga hydroelectric project, Maneri-bhali and Tehri high dam. The projects under construction are Vishnu Prayag hydropower project, Tehri dam second phase (1,000 MW), Koteshwar dam, Lakhwar-Vyasi, Loharinag-Pala, Pala-Maneri, Arakot Tyuni, Tyuni-Plau, Khartoli-Lumti, Jakhol-Sankari, Bowala-Nandprayag, Nandprayag-Langasu, Bhairavghati, Tamak-Lata, Sela-Urthig, Karmali, Pala-Tiloth, Hanol-Tyuni, Bhilangana, Vishnigad-Pipalkoti, Garba-Tavaghat and Jad Ganga (Department of Hydropower 2010).

The macro-hydroelectric projects in the region are not sustainable because of fragility and instability of mountain terrain. There are conflicts among the dam construction agencies, the local people and social workers about the construction of macro dams. To meet the energy need and to harness water resources optimally, construction of micro-hydropower projects is the need of the hour. Construction of micro-hydropower projects is suitable in the fragile mountain terrain and it will reduce the environmental degradation and rehabilitation problems. There are numerous suitable places, where these small projects can be commenced. The geologists and geographers should take initiative in selection of the potential sites for the construction of micro-hydropower projects keeping stability, settlement and agricultural land in consideration.

Some hydropower projects were case studied and a separate discussion was carried out on each project to understand the overall scenario. Table 6.2 shows that 13 hydropower projects of 3003 MW capacity are proposed or under construction in eight river valleys in Garhwal region affecting total 101 villages directly or indirectly.

Hydroelectricity Generation Through Water Mills

Community participation in production of hydroelectricity through installation of water mills with financial assistance from the government agencies is vital in this hilly region to meet the energy need and reduce firewood burden from the forest. The water mills are successfully commencing in many areas where the governmental agencies have provided financial assistance to the rural villagers to run this entrepreneur. The water mill electricity production is both sustainable and renewable in nature.

Many watermills—both improved and traditional were found in the study region. An improved watermill is constructed in the Khanda Gad sub-watershed with the financial assistance of the state government. This watermill is meeting the dual function of grain grinding and electricity generation. There are several other examples where the people are earning their livelihood through improved watermills.

Table 6.2 Details of selected hydropower projects

Hydropower projects	River valley	Capacity in MW (Construction company)	Number of affected villages and impact on them due to construction	Current situation
Vishnuprayag	Alaknanda	400 (JP)	5 villages, Perennial sources of water are dried up and cracks are placed in the villages. 30 houses damaged in Chai village	16 km tunnel constructed, connecting Lambagar and Chai village
Tapovan-Vishnuprayag	Dhauli Ganga	520 (JP)	5 villages, agricultural land, forestland and grazing land affected	Construction work has been started
Lata-Tapovan	Dhauli Ganga	162 (JP)	5 villages, agricultural land, forestland and grazing land affected	Construction work has been started
Singoli-Bhatwari	Mandakini	90 (L and T)	16 villages, agricultural, grazing and forestlands affected	Construction work has been started
Srinagar	Alaknanda	330 (JBK)	20 villages and Dhari Devi temple, Individual and community land affected	Construction work is in peak
Danawa-Churena	Balganga	5 (Gunsaula)	6 villages, irrigated agricultural land, grazing land, community land affected	Power house has been constructed
Bhilangana	Bhilangana	22 (Gunsaula)	4 villages, irrigated land and forestland	Powerhouse has been constructed and tunnel is under construction
Devlang (Ghuttu)	Bhilangana	24 (Gunsaula)	10 villages, forestland, community land and grazing land affected	Tunnel is being constructed
Maneri-Bhali Phase 2	Bhagirathi	304 (NHPC)	22 villages. Natural perennial water resources are dried up. Cracks are appeared in the houses	Under construction. Tunnel is likely to be completed
Pala-Maneri	Bhagirathi	480 (NHPC)	2 villages, fertile cultivable land and 6.8 ha forestland	Work started in 2007 but in 2008 the work is rigorously stopped due to people's agitation

(continued)

Table 6.2 (continued)

Hydropower projects	River valley	Capacity in MW (Construction company)	Number of affected villages and impact on them due to construction	Current situation
Lohari-Nag-Pala	Bhagirathi	600 (NHPC)	6 villages, 112 ha fertile land	-Do-
Jakhol-Sankari	Tons	33 (NHPC)	5 villages	Proposed, Peoples' agitation continued
Netwad-Mori	Yamuna	33 (NHPC)	5 villages	-Do-
Total 13	8 River valleys	3003 MW (5 companies)	101 villages	2 projects are proposed and 11 are under construction

Source Compiled by the author

Potentials of Micro-Hydroelectric Projects

The feasibility of micro-hydroelectric projects in mountain regions is considerably high where as macro dams have very limited feasibility. It is mainly due to the fragility and instability of mountain terrain. Macro-hydroelectric projects are not only infeasible from the stability point of view but they are unsustainable in terms of finance involved. The case of Asia's highest Tehri high dam is an example; it took nearly three and half decades to attain its present level. The expenditure for its construction doubled during the construction phase. A group of local people agitated against its construction and this agitation continued for nearly two decades. The economic viability and geological feasibility of micro-hydroelectric projects is considerably high, even a capacity of 2 × 2.5 MW power projects are quite suitable. The Government has already constructed few micro-hydroelectricity projects. A hydroelectricity project has been constructed in the confluence of the Kail Ganga and Pindar River, providing electricity to 25 villages of surrounding areas of Debal development block. Likewise, many other micro-hydroelectricity projects can be developed in this region.

References

Grotzbach, E. (1994). Hindu-Heiligtümer als Pilgerziele im Hochhimalaya. In: Erdkunde *48*(3). pp. 48.

Nepal Tourism Board. (2000). Welcome to Nepal. pp. 45–51. http://www.welcomenepal.com/Brochure.asp, 03/03/03.

Nielsen, AC ORG-MRG. (2007). Collection of tourism statistics for the state of Uttarakhand. Ministry of Tourism, Government of India.

Rangan, H. (2000). *Of Myths and Movements: Rewriting Chipko in Himalayan History. London: Verso. 70*, 72–74.

Sati, N. & Mansoori, I. K. (2009). *Pilgrimage tourism*. New Delhi: Mittal Publications

Sati, V. P. (2013). Cultivation of medicinal plants and its contribution to livelihood enhancement in the central Himalayan region. *Advancement of Medicinal Plant Research, A Net Journal*, *1*(2), 17–23.

Sati, V. P. (2004a). Systems of agricultural farming in the Uttaranchal Himalaya, India. *Journal of Mountain Science*, *2*(1), 76–85. http://www.imde.ac.cn/journal

Sati, V. P. (2004b). Systems of vertical horticulture in the Alaknanda Basin of Garhwal Himalaya. *Annals of Nagi*, *XXIV*(2), 49–58.

Sati, V. P. (2004c). Horticultural development in Hills: A case for the Alaknanda Basin. (Ph. D. Thesis). New Delhi: Mittal Publications. www.vedamsbooks.com/38930

Sati, V. P. (2004d). Uttaranchal: Dilemma of plenties and scarsities. New Delhi: Mittal Publications. www.vedamsbooks.com/38930

Walter, G. (2000). Tourism development and environmental management in Nepal: A study of Sagarmatha National Park and the Annapurna Conservation Area Project, with special reference to Upper Mustang. MSc Thesis, The International Centre for Tourism and Hospitality Management, Bournemouth University. pp. 9.

Chapter 7
Case Studies

Case studies of the two sub-watershed regions—the Khanda Gad sub watershed (KGSW) of the lower Alaknanda basin and the Kewer Gadhera sub-watershed (KGSW) of the Middle Pindar basin were carried out. Villages were the unit of detailed study and household level survey was conducted to know about the socio-economy of the region and to identify the major livelihood options that can be enhanced and diversified. Village wise data on population profile and land use pattern were gathered from the secondary sources, mainly from the COI for the years 1971 and 2001. Data were also gathered from the Patwari circles (a revenue unit) of 2007 for accessing the land use pattern. Primary data were gathered on farming systems and income of the people through formulation of structured questionnaire and personal interview of the households (HHs). Detailed illustrations of the case study villages in the two sub-watershed regions are as follows.

The Khanda Gad Sub-Watershed

The KGSW of the lower Alaknanda basin is located in the Southwest part of the Alaknanda River. This is a perennial stream, outlets from the '*Danda of Khirsu*' in Khirsu block of Pauri District, inlets into the Alaknanda River at opposite side of Kirtinagar town in the left bank. It is extended between 30° 9′ N and 30° 12′ N latitude and between 78° 45′ E and 78° 48′ E longitude. It is covered by dense forests, mainly pine forest in the lower elevation and oak forest in the higher reaches of the watershed. Besides, small bushes and shrubs are also found, mostly in the lower elevations before it meets with the Alaknanda River. Agricultural land is available in the valley regions and mid-slopes where traditional subsistence cereals are cultivated. Pauri town is a district-headquarter, located on the top of the watershed, facing to north-east direction. The other small service centres are Khanda, Khirsu and Chaubatiya. The valley regions remain hot during the summer while the highlands have feasible climatic conditions. However, during the winters, the valleys become severely cold and remain under dense fog for around three months.

V. P. Sati, *Towards Sustainable Livelihoods and Ecosystems in Mountain Regions*, Environmental Science and Engineering, DOI: 10.1007/978-3-319-03533-8_7,

Meanwhile, the highlands receive intense snowfall. The cold waves from the Great Himalayan range passes through the entire region and accentuates the frequency and intensity of this cold.

The studies on the cereal crops in the Himalayan region reveal that these crops dominate in the cropping pattern but, this crop cultivation do not support livelihoods at sustainable level. However, these crops ensure rich agro-biodiversity and they have the natural advantage related to the ecosystems of the region. During the field visits, the residents of Khanda Gad were interviewed. A resident, Mr. Kathait, has about 2 ha agricultural land, where he grows wheat, mustard and masure (lentil) during the *rabi* season followed by paddy, mandua (finger millet) and pulses during the *kharif* season. The production from the crops that he grows is very insufficient even to meet the daily food requirement. Therefore, to manage livelihoods he opened a shop in the village and now is able to meet the daily food requirements. This livelihood enhancing/diversifying trend is seen everywhere in the Himalayan region where the villagers are choosing to have cereal crop cultivation along with some other income generating activity. The agricultural land is fragmented and characterised by the narrow patches of terraced fields. Khanda is a small service centre, located on the right bank of the Kanda Gad, in the middle way to Srinagar town from Pauri. Small patches of agricultural fields are found along the sides of Khanda Gad and some of them are irrigated where intensive cultivation of wheat and paddy is carried out.

Remarkable changes have been observed in the cropping pattern. During the 1990s, five Nepali families immigrated to this region and the villagers leased out some agricultural land to these families. They started cultivating cash crops, mainly off-season vegetables—tomato, potato, onion, cauliflower, green vegetables and beans. Now, all families own 2 ha land. Due to adequate irrigation facilities from the perennial stream (Khanda Gad) and the use of composed manure, per ha yields of these cash crops are considerably high and they are adequate to ensure a sustainable livelihood. Production of off-season vegetables is high, for example in the case of tomato about 100 boxes of tomato (each box carries 18 kg) is available from 2 ha land. Market is easily available for sale of products, as this area is highly accessible and is connected to the national highway. Peas production is also high and in a season, growers are able to earn about Rs. 50,000 (IC).

It is interesting to note that even the potential of cultivating off-season vegetables is quite high in the study area; the area under cash crops is considerably less, whereas large areas under farming are still devoted to the cultivation of cereals and millets. There are 12 villages where these Nepali immigrants are cultivating cash crops and they are Margaon, Khanda Srikot, Margadna, Bhitai Malli, Kaldung, Dhanak, Gaduwa Gad, Kamand, Dov, Shiyar Malla, Rawat Gaon and Bhitai Talli. A study was carried out on the population profile, land use and cropping pattern of these 12 villages and the results are as follows.

Table 7.1 Area, HHs and changes

Village name	Area (ha)	HHs		
		1991	2001	Change in %
Margaon	9.3	35	31	−11.4
Shrikot Khanda	122.0	67	109	62.7
Margadna	32.7	18	8	−55.5
Bhitai Malli	73.4	71	76	7.04
Kaldung	80.9	31	32	3.2
Dhanak	26.1	19	27	42.1
Gaduwa Gad	44.7	11	10	−9.1
Kamand	25.5	17	23	35.3
Dov	53.7	26	24	−7.7
Shiyar Malla	33.8	0	–	0
Rawat Gaon	46.6	7	7	0
Bhitai Talli	71.9	20	21	5
Total	620.6	322	368	14.3

Source COI (1991, 2001)

Table 7.2 Population structure

Variables	Total	Population	Male		Female	
Population structure	1991	2001	1991	2001	1991	2001
Population (0–6)	269	172 (−36 %)	138	86 (−38 %)	131	86 (−34 %)
Total population	1580	1618 (2.4 %)	736	774 (5.1 %)	844	844 (No change)
Population density (person per ha)	3	3	2	2	1	1
Literacy	987 (62.5)	1228 (75.9)	567 (77)	695 (89.8)	420 (49.7)	533 (63.2)
SC population	399	392 (−1.7 %)	204	195 (−4.4 %)	195	197 (1.0 %)
Sex ratio	956	1,010				

Source COI (1991, 2001)

Population Profile 1991–2001

A population profiling was carried out in the 12 villages of the Khanda Gad. Table 7.1 shows area in hectares, number of HHs in 1991–2001 and changes in HHs numbers. There was a 14.3 % increase registered in the number of HHs in 12 villages. Some villages also recorded negative changes.

Table 7.2 tabulates population under 0–6 age group, total population, density of population, literacy rate, schedule caste (SC) population and sex ratio. Data from the two censuses were gathered (1991–2001). In 1991, the total population under 0–6 age group was 269, which were reduced to 172 in 2001. During this period, total 97 persons have been decreased (−36 %). Female and male population under

Table 7.3 Changes in land use (ha) 1971[a]–2007[b]

Village name	Total area	Forest land	Irrigated land	Unirrigated land
Margaon	−22.7	3.6	1.9	−12.3
Shrikot Khanda	−10.4	12.8	5.0	2.9
Margadna	5.9	–	0.4	6.7
Bhitai Malli	−0.6	5.2	3.7	28.5
Kaldung	−11.9	4.9	−0.1	35.2
Dhanak	0.5	3.0	0.2	6.9
Gaduwa Gad	−2.9	–	−0.5	17.1
Kamand	−0.3	5.5	−0.4	0.5
Dov	10.1	14.3	–	2.0
Shiyar Malla	20.6	–	1.3	−1.5
Rawat Gaon	−0.2	–	0.4	14.8
Bhitai Talli	−94.5	7.1	−0.7	−55.2
Total	−105.8	56.4	8.8	45.6

Source [a] COI (1971), [b] Patwari circles, Srinagar and Pauri

0–6 age group was decreased (−34 and −38 % respectively). There was only 2.4 % increase in total population during 1991–2001. Density of population was three persons per ha area (two male and one female). Literacy rate increased from 62.5 % in 1991 to 75.9 % in 2001. Male literacy was 89.8 % and female literacy was 63.2 %. Sex ratio was 1,010 female per 1,000 male.

Changes in Land-Use Pattern

Data on land-use pattern of 12 villages of the KGSW were gathered from the COI, 1971. The year 2007 data was gathered from the Patwari circle, Srinagar and Pauri. Three categories of land use such as forest cover, irrigated land and un-irrigated land (i.e. sown area) including total area of the villages were gathered and changes in land use from 1971 to 2007 were calculated. Table 7.3 shows land-use change.

For the last three decades, the total land area of the villages has recorded a decreasing trend. It was 726 ha in 1971, and in 2007 it receded to 620.2 ha (−105.8 ha). Contrary to this, forestland (community forest) increased from 2.4 ha in 1971 to 58.8 ha in 2007 (56.4 ha). Irrigated land also increased by 8.8 ha. The increase in forestland was mainly due to Forest Act of India (FAI) 1982, establishment of *van panchayat* (elected members of the village assembly who look after the conservation of forest) to ensure peoples' participation and because of the increasing awareness among the villagers towards conservation of forest. Increase in sown area was due to high growth in population. The rate of increase was considerably less as compared to the increase in population; this was because of large-scale emigration and land abandonment. The other categories of land such as wasteland and cultivable wasteland also recorded an increase.

Demographic Profile of Nepali Immigrants

About 13 years ago, some Nepali immigrants visited the villages of Khanda Gad and they temporary settled there and started cultivation of off-season vegetables. A case study was done during the period 2008–2009 on the demographic profile of Nepali immigrants and their production of off-season vegetables. The total number of HHs was 65, of the total population of 406 and sex ratio was 1,034. Further, literacy rate was only 31.3 % (among children, it was 90 %) with 45.3 % girl's literacy rate. Males and females were equally working on the farmlands.

The Nepali immigrants started cultivating cash crops such as potato, tomato, onion, cauliflower, capsicum, spinach, cucumber, pumpkin and beans at commercial level. They used both chemical fertilizer (Rs. 1,400/ha) and manure (Rs. 500/ha). Manure is easily available from the nearby villages. Every HHs has a pair of oxen, which they use to plow the fields and ensure manure availability. Responding to a question from the researcher, a head of Nepali resident in Dhanak village informed that there was no negative change in the production of off-season vegetables. To ensure high production these residents were using chemical fertilizers. However, still subsistence agricultural practices are carried on in the mid-altitudes of the sub-watershed, but the output from the traditionally grown cereals is considerably low.

Table 7.4 shows production and per ha yields from the cultivation of off-season vegetables and cereal crops in 2007–2008 in 12 villages of the KGSW. Cauliflower, cucumber, tomato, capsicum, beans and potato are the main off-season vegetables grown in these villages. Among cereal crops, wheat, rice, barley, millets, pulses and oilseeds are grown. Table 7.4 shows that per ha production of cucumber is highest (103.3) followed by tomato (89.5) and potato (53.3). Rice and barley (36.8 each) obtain the highest per ha yield followed by oilseeds (35.4). In a nutshell, per ha yield of off-season vegetables is 67.7, which is high in comparison to cereal crops (28.0). It shows that cultivating off-season vegetables is economically viable. If a sizeable proportion of cultivable land is devoted to cultivating cash crops, then food security can be ensured.

Table 7.5 shows area (ha) and annual income from the cultivation of off-season vegetables and cereals in 2007–2008. The total area under cultivation of off-season vegetables was 51.5 ha and annual income was Rs. 133,8100 (Rs. 25,982 per ha) while the area under cereal crops was 330.4 ha land and Rs. 502,1600 was the annual income, which means Rs. 15,198 per ha. Similarly, the number of households involved in the production off-season vegetables was 65, while 368 HHs were involved in the cultivation of cereal crops.

Table 7.4 Production and per ha yield from off-season vegetables and subsistence crops 2007–2008

Off-season vegetables	Area (ha)[a]	Production (in quintal)	Per ha yield	Subsistence crops	Area (ha)[a]	Production (in quintal)	Per ha yield
Cauliflower	15	727	48.5	Wheat	100	1,520	15.2
Cucumber	3	310	103.3	Rice	100	3,680	36.8
Tomato	20	1,790	89.5	Barley	50	1,840	36.8
Capsicum	3	118	39.3	Millets	50	1,472	29.4
Beans	3	144	48.0	Pulses	20	368	18.4
Potato	7.5	400	53.3	Oilseeds	10.4	368	35.4
Total	51.5	3,489	67.7	Total	330.4	9,248	28.0

Source Primary collection
[a] Patwari circles Srinagar and Pauri

Table 7.5 Area and annual income of off-season vegetables and cereal crops 2007–2008

Off-season vegetables	Area (ha)	Annual income (Rs.)	Cereal crops	Area (ha)	Annual income (Rs.)
Cauliflower	15	508,900	Wheat	100	1,472,000
Cucumber	3	217,000	Rice	100	2,520,000
Tomato	20	179,000	Barley	50	760,000
Capsicum	3	118,000	Millets	50	121,600
Beans	3	115,200	Pulses	20	70,000
Potato	7.5	200,000	Oilseed	10.4	78,000
Total	51.5	1,338,100	Total	330.4	5,021,600

Source Primary collection

The Kewer Gadhera Sub-Watershed

Kewer Gadhera is a sub-watershed of the Pindar basin. It originates from the root of Kanpurgarhi top (a highest point of the watershed) and confluences into the Pindar River at Narain Bagar. It is extended between 30° 5′ N and 30° 8′ N latitude and between 79° 20′ and 79° 23′ E Longitude; it flows in the middle watershed of the Pindar basin, it is a perennial stream. It has four sub-perennial tributaries. The watershed of stream makes steep to gentle slopes. During monsoon, it flows above denser marks due to heavy downpour. It flows below average during the winter season. Throughout its course, water is mostly unused for drinking as well as for irrigation purposes. It is because of rough, rugged and precipitous landscape of the watershed. Only in few places, small patches of terraced agricultural fields are irrigated. The whole watershed of the Kewer Gadhera is prone to soil erosion. Largely, during monsoon season, when entire watershed receives heavy downpour (mostly the winds blow from the south-east direction), landslides and debris flows are very common. There are many instances when the entire villages along with agricultural lands washed away and that led to heavy losses of lives and property. Musudiar landslide is an example, which took place in August 1993 and that led to death of 20 animals.

Until 2010, about 20 % villages were connected by the road network. Recently, roads are constructed and many villages have proper road connection. Although, there is a plan to connect all villages by road transport, yet, the pace of implementation is very slow. People have to walk miles to do their daily works. A small service centre Narainbagar is located on the bank of the Pindar River and on road head connecting Karnprayag and Gwaldom. This service centre fulfills the subsidiary needs of the villagers.

Subsistence cereal farming is the main occupation of the people. '*Barah Naja*' (12 grains), are the main crops grown, which include cereals, millets, pulses, and oilseeds. Agriculture is carried out on the narrow patches of terraced fields in the mid-latitudes and the highlands of the watershed. Some farm terraces are also available for cultivation of crops in the valleys. Loss of upper layer of soil and its fertility due to large-scale erosion has been resulted in low production and per ha yields of crops. Further, the modes of agricultural practices are traditional. Crops are grown in two different seasons. First season is *rabi* which starts from November–December and lasts to May–June. In this season, wheat is the main crop followed by barley and mustard. Earlier, barley was the prime crop. Second season is kharif. Rice, millets, maize, pulses and oilseeds are the main crops of this season. It begins from April and lasts in Oct. Changes in farming systems have been noticed during the last decades. A small proportion of land has been shifted to cultivation of paddy and wheat from the cereal crops. Still, in the highland villages such as Kimoli, Swan and Jhijodi, cereals are grown largely. In the other villages, which are located in the mid-altitudes and the valleys, rice and wheat along with millets are extensively grown.

Case study of 17 villages of the KGSW was carried out. The villages are Kewer Tall and Malla, Bhagoti, Ratni, Keshwan, Gadsseer, Bunga, Jhijodi, Ali, Leguna, Bedula, Chirona, Kaub (cluster of four villages), Naini, Swan Malla and Talla and Kimoli. Village wise study of land use and population profile was done.

Table 7.6 shows that revenue area of the villages remains more or less unchanged. It also varies from one village to another. In Jhijodi, village area reduced −42 ha between 1971 and 2007. This was followed by Keshwan (−24.5 ha). Swan Malla registered increase in land (26.2 ha) followed by Bunga (17.4 ha) and Gadseer (15.6 ha). The changes in the revenue area in the villages were mostly due to conversion of revenue land into forestland and vice versa. During this period, the population increase was negligible. It was due to exodus out-migration from these villages.

Table 7.7 shows the population profile in the villages of KGSW. There was 87.8 % population increase during the past several decades. Of it, 95.5 % was male population and 81.3 % female population. Density of population increased from 2 people/ha to 5 people/ha. Literacy rate increased from 27.8 % in 1971 to 61 % in 2001. Male literacy was higher (73.6 %) than female literacy (49.3 %). Increase in SC population was also registered during the same period.

Table 7.8 explains the land cover changes during 1971–2007. Community forestland shows an increase by 107.5 ha. Highest increase in forest cover (56.6 ha) was noticed in the Bunga village followed by Kaub (29.3 ha). Bedula

Table 7.6 Changes in area and HHs (1971–2007)

Villages	Area			HHs		
	1971[a] (ha)	2007[b]	Change	1971	2007	Change
Kewer Talla	32	31.1	−0.9	68	247	179
Kewer Malla	47.5	45.4	−2.1	13	26	13
Bhagoti	100.04	104.5	4.4	67	88	21
Ratni	37.6	37.8	0.2	15	27	12
Keshwan	50	25.5	−24.5	19	21	2
Gadseer	156	171.6	15.6	73	103	30
Bunga	112	129.4	17.4	60	70	10
Jhijodi	204	162	−42	116	150	24
Ali	11.5	11.1	−0.4	7	15	8
Leguna	14	19	5	9	20	11
Bedula	71.6	71	−0.4	20	36	16
Chirona	26.8	27	0.2	4	11	7
Kaub	231.6	228.6	−3	144	242	−2
Naini	22.8	22.6	−0.2	14	20	6
Swan Malla	2	28.2	26.2	13	16	13
Swan Talla	56.4	58.3	1.9		43	43
Kimoli	242	247.9	5.9	111	168	57
Total	1417.84	1,421	3.3	753	1,303	450

Source [a] Calculated from census, [b] Patwai circle

Table 7.7 Population profile (1971–2007)

Variables	Total population		Male		Female	
Population structure	1971	2007	1971	2007	1971	200
Population	3,521	6613 (87.8 %)	1622	3171 (95.5 %)	1,899	3442 (81.3 %)
Population density (person per ha)	2	5 (3 person)	1	2 (1 person)	1	2 (1 person)
Literacy %	27.8	61.0	48.0	73.6	10.5	49.3
SC population	900	1797 (99.6 %)	431	843 (95.6 %)	469	954 (103.4 %)
Sex ratio	1,170	1,085				

Source Calculated from census of 1971 and Patwati circle, 2007

village registered 18.1 ha land increase under forestland category. There are villages where forestland decreased. Irrigation is available only in the Kewer Talla village where a small portion of cultivable land is irrigated. Sown area decreased considerably. Due to emigration, agriculture land was kept abandoned and forestland had increased.

Some other areas were also studied to observe livelihood pattern. These are Shubhas Herbal Nursery, Janglechatti Village and Takori Gad catchment and their detailed description is given below.

Table 7.8 Land cover changes in the KGSW between 1971[a] and 2007[b]

Village name	Total area	Forest area	Irrigated	Un-irrigated
Kewer Talla	−0.9	−1.2	−0.8	0.5
Kewer Malla	−2.1	−1.2	–	−0.1
Bhagoti	4.5	1.2	–	24.8
Ratni	0.2	1.7	–	0.4
Keshwan	24.5	–	–	16.8
Gadseer	15.6	3.5	–	1.3
Bunga	17.4	56.6	–	4.2
Jhijodi	−42	−29.3	–	26.3
Ali	−0.4	5.4	–	−0.1
Leguna	5	5	–	−1.9
Bedula	−0.4	18.1	–	−0.3
Chirona	0.2	6.7	–	0.1
Kaub	3	29.7	–	2.9
Naini	−0.2	−0.1	–	−0.8
Swan Malla	26.2	−0.8	–	3.7
Swan Talla	1.9	5.8	–	1.4
Kimoli	5.9	6.4	–	13.9
Total	58.4	107.5	−0.8	93.1

Source [a] Census handbook 1971, [b] Patwari circle, Narainbagar

Shubhas Herbal Nursery

Shubhas Herbal Nursery (SHN) is located about 2 km away from the Gwaldom town towards Talwadi at a height of 1,900 m. It is spread on about 5 ha land. Shri Jodh Singh Badiyari is the owner of this farmland. High quality and quantity of apple were grown before the 1990s and about 50-quintal apple were produced in a season. Apple cultivation is hindered by climate change and as a result, several diseases in apple trees appeared. This farmland has now been changed for cultivating medicinal plants. A herbal plant (German Made) was established to process the herbs for making medicinal products (cost of establishment was Rs. 700,000). A poly-house was built with the help of the state government. The main herb is *Geranium*, which is used for making medicines, scents and soaps. One litre of herbal oil costs Rs. 4,000. The owner was able to produce 25-liter oil in a season. A unique variety of rose with white flower (*Nurjahan Gulab*) is also grown largely in this farmland. Except this herbal plant, seasonal vegetables such as pumpkin, cucumber and potato are also cultivated with the use of organic fertilizer. About 250-quintal potato is grown in a season.

Junglechatti Village

Janglechatti village is located at 1900 m altitude in Chamoli District. There are about 300 HHs out of it, six HHs have out-migrated. The villagers are working in the agricultural fields. The total cultivable land is 100 ha. The agro-climatic

Table 7.9 Employment status, farming community and emigration

Village	Employment status				Farming community		Emigration		Total
	Teaching	%	Other	%	No.	%	No.	%	
Kaparniya	13	10	40	30	65	50	12	9	130
Bachwar	02	2	25	33	40	54	7	9	74
Jakholi	15	25	20	34	20	34	03	5	58
Bazeera	04	1	85	38	120	55	09	4	218
Dhankurali	00	0	20	429	25	53	02	4	47
Mayali	25	20	55	44	30	24	15	12	125
Baman Gaon	22	26	30	35	12	14	20	23	84
Barsir	09	5	60	39	70	46	12	7	151
Jainti	21	16	45	34	50	38	15	11	131
Serain	01	2	15	44	10	29	08	23	34
Total	112	10	395	37	442	42	103	9	1,052

Source Primary collection

conditions are quite feasible for growing all types of cereals, fruits and vegetables. Here, the Government has established a 'fruit belt' and a 'tea nursery'. The whole area is devoted for apple cultivation. A considerable proportion of land, which was otherwise not fit for the production of cereal crops, was devoted for this purpose. Tremendous success was achieved by the farmers in growing apple and still today, the farmers are able to grow considerable quantity of apple. Tea production is also increasing. The gentle slopes of forest and community land are reserved for tea cultivation. The future prospect of tea cultivation seems to be promising.

The Takori Sub-Watershed

Eleven villages—Dang Choura, Dugadda, Malu Pani, Semala, Takoli, Tolu, Tatyamandal, Maikhand, Jakhand, Amoli and Kandikhal of the Takori sub-watershed were case studied. All villages are located on the road head of Srinagar-New Tehri road. The agro-climatic conditions in these villages are feasible for producing paddy, wheat, turmeric, ginger, garlic, potato and onion crops. Pulses and oilseeds are largely grown. Recently, the cultivation of potato and ginger has got momentum and the people are quite interested to cultivate these crops.

The Lastar Gad Watershed

The landscape of Lastar Gad watershed is very much feasible for development of land-based resources. These include intensive agriculture, horticulture, herb culture, apiculture, non-timber based forest products and dairy farming. Eco-tourism and construction of micro-hydroelectricity has high potential. High soil fertility, dense temperate and pine forest, feasible environmental conditions, extensive agricultural fields with gentle slopes and availability of workforce characterise this watershed.

Table 7.9 shows employment status, farming community and emigration in the Lastar Gad sub-watershed of Mandakini basin. About 10 % population is engaged in teaching occupation, and about 37 % population is engaged in the other services. About 42 % people are engaged in agriculture. Out of the total population, about 9 % people have out-migrated.

Conclusions

Only cereal crops cultivation does not ensure the daily food requirement of the populace. Increased population pressure on the limited terraced agricultural fields has further accelerated food scarcity. This has led to the transformation of cereal crops into cash crops cultivation so as to ensure food security. Similarly, large-scale emigration towards metropolitan cities and the plains of Ganges in search of better livelihood options is taking place since the last several decades. The case study villages have favourable conditions for achieving food security through cultivating cash crops. The following suggestions are given to enhance livelihoods and to achieve food security: (1) a sizeable proportion of land should be transferred from traditional crops to cash crops cultivation; (2) millets cereal farming should be practiced parallel to maintain agro-biodiversity; (3) rearing of livestock for producing milk and milk products; and (4) community participation should be ensured.

Chapter 8
Mountain Ecosystems

Mountains lie in every continent of the world. The Himalaya, the Andes, the Alps, the Rockies, the Atlas, the Hindkush, the Suleman, the Kirther and the Altaintag are the major mountain ranges. Similarly, there are many other well-known highlands such as the Elburz Mountains in the Islamic Republic of Iran, the Cairngorms in Scotland and the Fouta Djallon Highlands in the West Africa. These mountains are characterised by massive global diversity—from the tropical rain forests to the permanent ice and snow. Climate—temperature, rainfall and humidity changes from the valleys to the highlands and from one mountain range to other because of variations in altitude, slope aspects and nearness from the snow clad regions. Most of the mountains of the world are elevated above 4,000 m and covered by snow throughout the year.

Mountains are complex and fragile ecosystems with marked topography, highly differentiated climatic conditions and vertical processes. They are the water towers of the world—providing freshwater to at least half of the world's people for domestic use, irrigation, industry and hydropower—and storehouses of global biodiversity. However, mountains are also high-risk environments; avalanches, debris-flow, landslides, mass movement, flashfloods, volcanic eruptions, earthquakes and glacial lake outburst floods threaten life in mountain regions and surrounding areas, while fragile soils and vegetation cover make mountain areas vulnerable to environmental degradation. Mountains can be defined in various ways. The most commonly used definition of mountains was developed by the United Nations Environment Programme (UNEP) World Conservation Monitoring Centre. It divides mountain areas into seven classes, based on altitude, slope and the resulting environmental gradients:

Class 1	Elevation > 4,500 m
Class 2	Elevation 3,500–4,500 m
Class 3	Elevation 2,500–3,500 m
Class 4	Elevation 1,500–2,500 m and slope 2°
Class 5	Elevation 1,000–1,500 m and slope 5° or local elevation range (7 km radius) > 300 m
Class 6	Elevation 300–1,000 m and local elevation range (7 km radius) > 300 m
Class 7	Isolated inner basins and plateaus less than 25 km^2 in extent that are surrounded by mountains but do not themselves meet any of criteria 1–6

V. P. Sati, *Towards Sustainable Livelihoods and Ecosystems in Mountain Regions*, Environmental Science and Engineering, DOI: 10.1007/978-3-319-03533-8_8,

According to this definition, the global mountain area is almost 40 million km^2. Mountains play important role in influencing global and regional climates and weather conditions. By intercepting the global circulation of air, they have a decisive effect on wind, precipitation and temperature patterns. Mountain climates vary considerably—from year to year, season to season and day to day, at different altitudes and on slopes with different exposures. Traditional land use systems utilise these small-scale variations, and mountain people have developed efficient techniques for farming, livestock breeding, forestry and water use that are adapted to life on steep slopes and in harsh, unpredictable conditions. In many mountain areas resources are limited. Mountain dwellers make the most of it by growing sun-loving plants on the warmest slopes and moving livestock to graze on high summer pastures after the snow has melted. Diversified and well-adapted traditional land use systems protect the soil from erosion, conserve water and maintain a rich biodiversity. Price and Kim state that 'given the very different characteristics of the world's diverse mountain regions, even on one continent, it is probably best not to propose a precise definition of sustainable mountain development, but to recognise that it is a region specific process of sustainable development that concerns both mountain regions and population living downstream or otherwise dependent on these regions in various ways'. Sustainable mountain development requires that mountain ecosystems be managed in ways that allow providing goods and services for local livelihoods and lowland people both in present and in future times. By addressing environmental, economic, social, cultural and political issues in a holistic manner, sustainable mountain development aims at improving the lives of mountain people and the life-support systems of the surrounding lowlands.

Himalayan Ecosystem

The Himalayan ranges are the youngest and loftiest among the mountain systems of the world. They represent a highly complex and diversified system both in terms of biological and physical attributes. The region has discrete geographic and ecological features. It produces a distinctive climate of its own and influences the climate. However, the variations in topographical features along three-dimensional framework cause diversity in climate and habitat conditions within the region.

The Himalayan landscape systems are unique. These systems, with their steep slopes and sharp gradients, are heterogeneous and exhibit sharp and most often systemic changes in climatic variables over very short distances; consequently they result into enhanced changes in hydrological processes, with accelerated direct runoff and erosion. Major rivers of the region have their origin from these mountains and are the source of water for a large proportion of the human population within and outside the mountain region. Many of the world's crops originate in mountains, a crucial resource that should be conserved for sustaining modern agriculture. Natural wealth in this region, including geological assets, forms an important part of the Himalayan ecosystem. All this has contributed to a whole range of diversity in indigenous human habitations, cultures and knowledge systems. The region is largely inhabited by indigenous societies. Living in biodiversity rich areas of the country, the mountain people are dependent upon biodiversity for meeting with their livelihood needs. Therefore, sustaining biodiversity in the region also means protecting the interests of this people. The region serves as a rich repository of plant and animal wealth in diverse ecological systems. These ecosystems reflect a mosaic of biotic communities at various spatial and organisational levels. Recognition of the Himalaya as one among 12 global biodiversity hotspots aptly reflects its wide-ranging ecological significance.

The vulnerability of the biological and physical features of the Himalayan ecosystem towards natural and human induced disturbances is well recognised. Immediate actions are required to ensure sustenance and protection of the ecosystem. Among others, this ecosystem needs to be protected against both improperly conceived developmental activities and consequences of climate change need to be assessed. Further, in view of an inevitable need for conservation of biological diversity in a nutshell, sustainability of Himalayan ecosystems has gained paramount importance. Hence, the National Mission for Sustaining Himalayan Ecosystems (NMSHE) is considered as an important part of India's National Action Plan on Climate Change (NAPCC).

The NAPCC has enunciated the launch of a National Mission for Sustaining the Himalayan Ecosystem. The Mission is to deliver better understanding of the coupling between the Himalayan ecosystem and the climate factors. It is also to provide inputs for Himalayan sustainable development. This requires the joint effort of climatologists, glaciologists and other experts. Exchange of information with the South Asian countries and countries sharing the Himalayan ecology is also required. There is a need to establish an observational and monitoring network for the Himalayan environment to assess freshwater resources and health of the ecosystem. The mission attempts to address some important issues concerning:

- Himalayan glaciers and the associated hydrological consequences,
- Biodiversity conservation and protection,
- Wildlife conservation and protection,
- Traditional knowledge societies and their livelihood and
- Planning for sustaining of the Himalayan ecosystem.

Recognising the importance of scientific and technological inputs required for sustaining the fragile Himalayan ecosystem, the Ministry of Science and Technology has been entrusted with the nodal responsibility of coordinating this mission. However, the mission requires valuable cooperation of Indian Himalayan States, the Planning Commission and the Ministry of Environment and Forests to achieve its goals. Sustainability of an ecosystem demands a balance and equilibrium among various forms of life and their surroundings established over long periods of time. Therefore, the mission recognises the need for creation and building of national capacities to observe and respond to changes in a sustainable manner. The mission proposes a coordinated effort in identification and strengthening of institutions engaged already in the conservation and management of the natural resources in the ICHR. It plans to effectively network such institutions with common and shared objectives, co-sharing of resources and co-generation of processes leading to ecologically sustainable development. Sustaining the Himalayan ecosystem as a national mission, it focuses on the rapid generation of four types of national capacities, they deal with: (i) Human and knowledge capacities; (ii) Institutional capacities; (iii) Capacities for evidence-based policy building and governance and (iv) Continuous self-learning for balancing between forces of nature and actions of mankind.

The primary objective of the mission is to develop in a time bound manner a sustainable National capacity to continuously assess the health status of the Himalayan ecosystem and enable policy bodies in their policy-formulation functions and assist States in the ICHR with their implementation of actions selected for sustainable development. This would call for suitable learning systems for balancing changes in ecosystem with responsible human actions.

Increasing temperatures and changing precipitation patterns in the ICHR can be expected to influence even more profoundly the regional mountain ecosystems and some sections of human population. The changes can be expected to be complex. Effects of climatic changes on the environment and people's livelihoods could impact health, agriculture, forests, water resources, coastal areas, species and natural areas. There is a compelling need for a concerted effort to understand the implications of climate change on vulnerable sections of people and develop special strategies for sustaining fragile ecosystems on which large number of people depend. In the above context, the National Environment Policy 2006 envisages a few measures for conserving the mountain ecosystem. Specifically, the policy proposes to:

- Adopt appropriate land use planning and watershed management practices for sustainable development of mountain ecosystem.
- Adopt 'best practice' norms for infrastructure construction in mountain regions to avoid or minimise damage to sensitive ecosystems and despoiling of landscapes.
- Encourage cultivation of traditional varieties of crops and horticulture by promotion of organic farming, enabling farmers to realise a price premium.
- Promote sustainable tourism through adoption of 'best practice' norms of eco-friendly and responsible tourism, creation of appropriate facilities and access to ecological resources and multi-stakeholder partnerships to enable local communities to gain livelihoods, while leveraging financial, technical and managerial capacities of investors.
- Take measures to regulate tourist inflows into mountain regions to ensure that these remain within the carrying capacity of the mountain ecology.
- Consider unique mountain landscapes as entities with 'Incomparable Values', in developing strategies for their protection.

Indian Himalayan Region

The Indian Himalayan Region (IHR) covers about 5.3 lakh km^2 geographical area and comprises the vast mountain range extending over 2,400 km in length between the Indus and the Brahmaputra river systems. Elevated from the low-lying river valleys to over 8000 m, it is around 300 km wide. The Himalaya is among the world's highest mountain chain and is characterised by a complex geologic structure, snowcapped peaks, large valley glaciers, deep river gorges and having rich biodiversity. Its ecosystem is vital to the ecological security of the Indian landmass and occupies the strategic position of entire northern boundary (North-West to North-East) of the country. Apart from the national security standpoint, IHR is also important for its high forest cover. More than 41.5 % of its geographical area is under forests, representing one-third of the total forest cover in India and 47 % of the 'very good' forest cover category. These forests are bases of goods and services. However, they are complex systems of climatic and geological processes. Destructive patterns of resource use and economic marginalisation have led to the situation of heavy resource degradation and associated environmental consequences on the highly diverse and fragile Himalayan ecosystem.

Effect of Climate Change on the Himalayan Ecosystem

The Himalaya is ecologically sensitive mountainous area, prone to adverse impacts of global climate changes on account of both natural causes and anthropogenic emissions. Further, its ecosystem resources are critical on the face of natural disturbances, anthropogenic activities and climate change. It has important implications for the formulation of management strategies and sustenance of dependent human societies. Some of the significant consequences arising out of the global warming on the Himalayan region could relate to (i) variability in the volumetric flow of water in the rivers; (ii) loss in biodiversity; (iii) unsustainable changes in ecology; (iv) glacier recession; (v) deforestation and degradation; (vi) conditions for impending natural disasters and (vii) dislocation of traditional societies dependent vulnerably on the Himalayan ecosystem.

The Ganges, the Brahmaputra, the Sind and other major river systems originate from the Himalaya. Any changes in the Himalayan glacier dynamics and melting are expected to severely affect about 1.3 billion people. The IHR is repository for about 8,000 species of flowering plants, i.e. nearly 50 % of the total flowering plants of India. Of this, nearly 30 % are endemic to the region. There are over 816 tree species, 675 edibles and nearly 1,740 species of medicinal plants in the IHR. The Himalaya with its vast green cover acts as 'sink' for carbon dioxide. Annual carbon sequestration by the forests of western and northeastern Himalaya is just one of the important ecosystem services being performed by the Himalayan forests. This service needs to be strengthened and exploited for global good. Further, about 17 % area under permanent snow cover and glaciers and about 30–40 % under seasonal snow cover form a unique water reservoir. This feeds several important perennial rivers that provide water for drinking, irrigation and hydro-power. The IHR is home to nearly 4 % of the country's population and is provider for their livelihoods.

The IHR is highly vulnerable due to geological reasons and anthropogenic challenges. These effects may well be exacerbated due to impact of climate change. Climate change is likely to adversely impact the Himalayan ecosystem through increased temperature, altered precipitation patterns, episodes of draught and biotic influences. This would not only impact the very sustenance of the indigenous communities in uplands but also the life of downstream dwellers across the country and beyond. The task force on mountain ecosystem for environment and forest sector constituted by the Planning Commission, Government of India, has prepared an account of State of Art in knowledge and institutional capacities associated with the Himalayan ecosystem.

Global Sustainable Development Depends on Mountain Resources

Mountains provide 60–80 % of the world's freshwater resources for drinking, for the lowland irrigation vital to food security, for industry and for hydropower, the most important source of green energy, contributing 20 % of global electricity generation. In drier areas, such as the Middle East, Central Asia, South Asia and parts of Western North America, virtually all freshwater is sourced out from the mountain rages. Mountains also supply important minerals and the genetic resources of major food crops. A total of 17 of the 34 global biodiversity hotspots are in mountains. One-third of all protected areas are in mountains; many of the world's largest cities including New York, Jakarta, Tokyo, Mumbai, Nairobi, Mexico City and Bogota rely on such areas for their water supply. In an increasingly urbanised world, mountains are the major destinations for recreation and tourism. The European Alps, with over 540 million overnights per year, are the second most important tourist destination worldwide after the Mediterranean coast. Tourism is also an increasingly important source of growth and employment in many mountain areas of developing countries.

Challenges to Sustainable Mountain Development

Mountains worldwide are facing severe challenges in terms of degradation of landscape and depletion of biodiversity resources. This leads to minimisation of goods and services that mountains provide to the local people inhabiting in or vicinity of mountainous regions. Land degradation due to inappropriate farming practices and extractive industries (mining) is a major problem, exacerbated by population growth and ambiguous resource governance and land tenure rights. About 90 % of the global mountain population lives in the transition and developing countries. Close to 30 % of them, or 300 million people, are vulnerable to food insecurity and many of them live in extreme poverty. Access to basic infrastructure such as health and education is often poor and this increasing the marginalisation of mountain communities. Mountains are particularly susceptible to environmental hazards, causing disasters, such as devastating floods, landslides or avalanches leading to loss of life, property and livelihoods. Many mountain areas are located in tectonically active zones with an increased probability of earthquakes and volcanic eruptions. At a global scale, 55 % of mountain land is susceptible to earthquakes, as opposed to 36 % of non-mountain land. Mountains are also very sensitive to climate change, as shown by the rapid melting of glaciers worldwide and the consequences reach far beyond mountain regions.

Chapter 9
Sustainable Mountain Development: Challenges and Opportunities

The world is currently facing a multitude of global challenge. Climate change and increasing natural disasters, food and energy crises, a growing population (predicted to reach 9.2 billion people by 2050), increasing water scarcity and desertification, deteriorating water quality, loss of biodiversity, overall ecosystem degradation, financial crises, political instability, migration and the growth of cities are some important challenges. These challenges disproportionally affect mountain regions and their inhabitants, particularly in the developing countries. Moreover, owing to the relevance of mountain goods and services, changes occurring in mountain areas reach far beyond the mountains. The global importance of mountains as water towers, biodiversity hotspots, indicators of climate change and hubs of traditional indigenous knowledge means that mountain regions also offer strategic opportunities to ensure sustainable development. Sustainable mountain development is thus key requirement to address these global challenges and to overall sustainable development.

Sustainable mountain development is essential for achieving food security, poverty alleviation and sustainable development. If poverty reduction can be achieved by sustainable development in theoretical aspects, the conceptual approach of sustainable livelihoods is suitable to hold up sustainability and could be operated as one of goals for poverty reduction in practical. Here sustainability is regarded as a function to preserve and maintain livelihoods by different means of utilisation various assets and capabilities. Sustainability should be more people-centre, which can be understood as the ability of people to maintain productivity even when facing unpredictable stresses and shocks.

Major Issues of Sustainable Mountain Development

Mountains are of strategic importance in response to global challenges. Migration is becoming an unprecedented problem at a large scale. Many mountain regions are faced with labour out migration, mainly of young men, whose remittances to their homes change local economies, livelihood options and social structures.

V. P. Sati, *Towards Sustainable Livelihoods and Ecosystems in Mountain Regions*, Environmental Science and Engineering, DOI: 10.1007/978-3-319-03533-8_9,

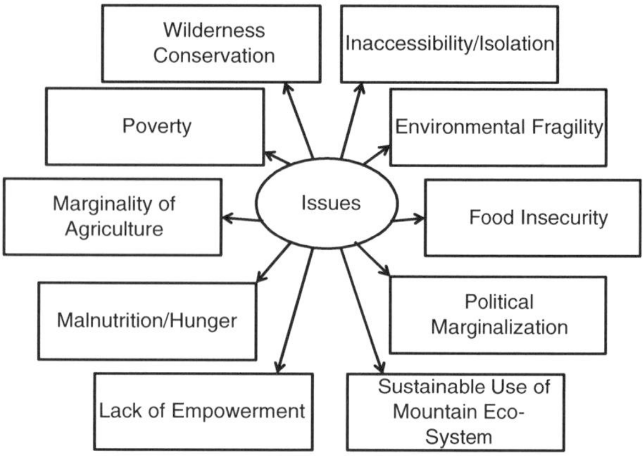

Fig. 9.1 Major issues of sustainable livelihoods in mountainous region

Although the vast majority of mountain people live in rural settings, yet urbanisation has reached 60–80 % in some areas of the Andes.

Climate change will compromise the role of mountain ecosystems as the world's water towers. Decreasing water flow from mountains will seriously affect agricultural production and food security, not only for mountain communities but also for the millions of people who live in lowland areas. It will also threaten the supply of water to large urban centres in the lowlands and the production of hydropower. In arid and semi-arid areas, which are particularly dependent on mountain water, increased water scarcity may even lead to conflict. Water shortage has also been associated with a decline in water quality, which may increase the risk of water-borne diseases. Investing in the sustainable management of mountain regions and the careful, wise and efficient use of mountain water is a global priority in a world heading towards a water crisis.

The demand for goods and services from mountains has grown considerably and will continue to grow further. Particularly in the context of the green economy, new opportunities for investment by both the private and public sectors are emerging, for example, in renewable energy and ecosystem services. This offers not only scope for economic development, but also puts increasing pressure on already fragile environments and scarce resources (Figs. 9.1 and 9.2).

The implementation of institutional arrangements that has the responsibility of the balanced development of social, ecological and economic capital is essential to ensure that new opportunities bring benefits and do not perpetuate the degradation of mountain socio-ecological systems. Suitable tools and methods that allow the valuation of mountain goods and services and adequate compensation mechanisms have to be put in place. To respond to global challenges, justice is needed to be ensured in the strategic importance of mountains. To achieve sustainable mountain development, there is need to invest in enhancing the knowledge base about mountain ecosystems' functioning and interactions with lowland areas.

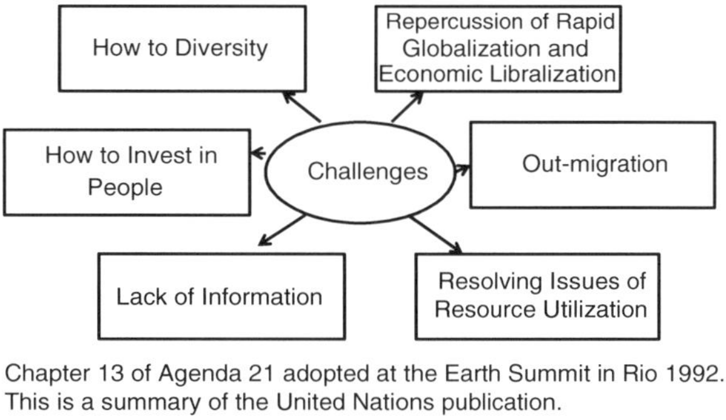

Fig. 9.2 Major challenges towards sustainable livelihoods in mountainous regions

Challenges to Sustainable Mountain Development

Climate change, increasing natural disasters, food/energy crises, population growth, water scarcity/desertification, loss of biodiversity, degradation of ecosystems, migration and growth of towns/cities—the planet is currently facing a multitude of challenges. Mountain regions and their inhabitants are not only disproportionally affected, but also offer significant opportunities for solutions. By providing key environmental services such as freshwater, biodiversity conservation and hydropower to more than half of humanity, mountain ecosystems play critical role in world development. Mountain systems are essential building blocks for long-term sustainable global development, poverty alleviation and the transition to a green economy. Mountain people, who are among the world's poorest and hungriest, are key to maintaining mountain ecosystems and their role in providing environmental services to downstream communities is inevitable. Mountain communities need to be empowered and their livelihoods improved, to enable them to take responsibility for the preservation of natural resources and to fulfil their role as mountain stewards. In spite of the obvious importance of mountain areas, sustainable mountain development does not receive the attention and priority it deserves. Investing in sustainable mountain development is a global priority for addressing the current challenges. It reaches far beyond monetary terms to embrace increased attention to and support in all aspects of mountain ecology and society.

Harsh climatic-environmental conditions, remoteness and often difficult access hamper development of the mountain regions. Mountain people frequently face political, social and economic marginalisation and lack access to such basic services such as health/education. Mountain communities live far from the centres of commerce and power, so they have little influence on the policies and decisions that affect their lives, and their voices often go unheard. Current global challenges such as climate change, economic developments and population growth exacerbate the hardships they face. Urbanisation and migration processes have a significant impact on mountain environments and societies. The fragility of mountain ecosystems makes the impacts of unsustainable development more severe and more difficult to correct than in other areas of the world. Sustainable approaches to development are therefore particularly important in mountain regions.

Recent environmental, economic and social developments have made many mountain regions increasingly disaster-prone. Disasters in mountains not only harm mountain communities but also have great impacts on the downstream areas, affecting millions of people. Mountains are greatly affected by destructive natural processes. Shifting tectonic plates cause earthquakes and volcanic eruptions, while heavy rains and snow on steep slopes produce avalanches, landslides, debris and mud flows and flashfloods. When combined with human interventions, particularly the construction of infrastructure and settlements in hazardous areas, such events turn into disasters causing damage, destruction, injury and death. Over the generations, mountain people have learned how to live with the threat of natural hazards and have developed well-adapted and risk-resilient land use systems. However, there is growing evidence that many mountain regions have become increasingly disaster-prone over the past few decades. Recent developments have significantly reduced mountain communities' resilience to natural hazards: population growth, the expansion of commercial agriculture and settlements and increasing urbanisation have reduced the availability of arable land; impoverished farmers are increasingly forced to clear marginal lands on steep slopes and reduce fallow periods; and more and more animals are grazing on mountain pastures and forested land.

Poor construction of mountain roads increases the likelihood of soil erosion and is one of the most important human-induced triggers of landslides. Roads tend to modify existing patterns of surface and groundwater drainage and increase the potential for instability. High rates of surface erosion increase sediment production. On roads that lack drainage structures, surface erosion increases by about 40 %. Sediments can greatly alter stream habitat and water quality and thus have serious effects on the downstream. Greater investments in road construction and restoration, improved road design and better maintenance practices are needed to limit the negative impacts of mountain roads. It is particularly important to stabilise the slope above/below the road cut; road design should include adequate paving, more numerous drainage channels, vegetative filter strips, out-sloping of the road surface to disperse runoff and narrower road surfaces to reduce the road tread. Roads should be located outside riparian areas, and care should be taken to avoid hazardous terrain and protect sensitive slopes.

Inappropriate farming practices and the destruction of mountain forests aggravate the pressure on fragile mountain ecosystems—ground cover is destroyed, soils are compacted and erosion accelerates, reducing soil fertility and increasing the potential for natural hazards. To participate in a wider market economy, mountain communities have had to alter their agricultural practices. Many mountain farmers have abandoned their traditional and diversified agricultural systems and rely increasingly up on a single cash crop for their livelihoods. Indigenous knowledge about local foods and agricultural practices has been eroded which led to a decline in the agricultural diversity. Such unsustainable development patterns exacerbate the pressure on natural resources and it aggregates the problems of food insecurity and malnutrition in mountain areas. The mountain livelihoods lose their resilience and become increasingly vulnerable to external economic shocks. The increasing exploitation of mountain areas by outside forces such as commercial agriculture, logging, mining and tourism enterprises puts additional dangerous pressure on these fragile ecosystems. Dams and roads can be hazardous if they are not properly constructed and managed.

Opportunities for Sustainable Mountain Development

Mountains play a decisive role in collecting and storing the most precious and threatened element for life on earth, i.e. freshwater. Rivers originating in mountains are living bonds that connect mountain to the lowland communities and provide water for irrigation, food production/domestic use. Mountains have approximately one-quarter of global biodiversity and are often rich in endemic species—animals and plants that can not be found in any other region. Mountains' altitudinal gradients, differentiated topography, changes in exposition over short distances, varying geology and soils offer a great diversity of habitats. Most of the world's major food crops, such as wheat, maize, potatoes and a large portion of domestic animals are found in the mountain areas. Mountain forests provide essential environmental goods and services such as timber, firewood, fodder, carbon storage and non-timber forest products. They capture and store precipitation, regulate surface and groundwater flows; they also ensure high water quality, ensure protection against natural hazards. Their diverse ground cover and highly developed root systems stabilise steep slopes and protect the soil from erosion. Mountains are a key source of energy, including biomass fuels, such as wood, and other forms of renewable energy. Their marked altitudinal gradients, exposure to air circulation and high solar radiation provide solar/wind/hydropower. Mountain regions and communities offer a wide variety of high-quality products such as organic food, beverages, handicrafts, herbs and medicinal plants. Mountain areas also provide raw materials such as timber, minerals and metals. Throughout history, mountain people have tended to settle in individual valleys, and many mountain regions have been areas of retreat or refuge for indigenous people, ethnic minorities and other communities that have been forced to move. The need to adapt to highly differentiated, fragile and inhospitable ecosystems has created a huge variety of indigenous experiences and knowledge. Mountain people are known for their unique traditions and practices, which contribute significantly to ethnic, cultural, linguistic and religious diversity. With more than 50 million visitors per year, mountains are some of the world's most important destinations for tourism. Beautiful landscapes, sporting/recreational possibilities, the unique traditions and cultures/lifestyles of mountain people attract increasing numbers of visitors, mainly from lowland cities. Tourism can be found in almost every mountain region and often dominates local economies, despite its seasonal character.

Throughout the world, mountains often embody and reflect people's religious and cultural values and aspirations. For millions of Buddhists and Hindus, the remote Himalayan peak of Mount Kailas points the way to the realm of the gods and the highest level of meditation. Mounts Sinai and Zion are places of worship in Judaeo-Christian tradition, while the Hopi and Navajo people perceive the San Francisco Peaks of Arizona as a divine source of the water on which they depend. For many people throughout the world, Mount Everest is an inspiring symbol of the ultimate. Mountains are perceived as the sites of revelation, inspiration, renewal and spiritual transformation, the sources of blessings such as water, life, fertility and healing and the abodes of ancestors and the dead. Their association with deities and evil spirits makes them places of worship, ceremonial sacrifice and pilgrimage.

The forces that shaped the world's mountains also made them rich in minerals as well as metals, including gold, copper, iron, silver and zinc. Owing to increasing demand, mines are now being opened even in remote mountain areas, particularly in the developing countries. Mining can bring large benefits, but it can also be devastating to fragile mountain ecosystems and local cultures, destroying the livelihood base of mountain communities. Massive quantities of waste, surface dumps and slag heaps are only the most visible consequences. Mining leads to atmospheric pollution and the loss of biodiversity and vegetative cover, which in turn destabilises mountain slopes. Water contamination is especially serious because mountains are reservoir of drinking/irrigation water. In some mountain regions of Africa where mines are located, arsenic levels in water are 1,000 times the accepted standard.

How Is Climate Change Affecting Mountain Areas?

Amongst different terrains, the mountains are most affected by climate change. Impacts such as decreasing water flows from mountains will have serious implications for the livelihoods of both the mountain and the downstream communities. Climate change is one of the most important global challenges affecting mountain ecosystems. Mountains host the most visible and sensitive indicators of climate change, viz. the melting of glaciers; and many scientists believe that the changes occurring in mountain ecosystems may provide an early glimpse of what could happen in lowland environments. Mountains can therefore be considered as early warning systems.

The Intergovernmental Panel on Climate Change (IPCC) predicts that by the end of the twenty-first century, human-incurred greenhouse gas emissions will lead to average global warming of between 1.1 and 6.4 °C. This will inevitably change the hydrological cycle, particularly, alter precipitation and runoff patterns. The greatest impacts are expected in areas where snowmelt dominates the hydrology; the melting and disappearance of glaciers, reduced snow storage and earlier snowmelt have direct effects on the amount and seasonality of runoff. For more than 1 billion people who currently rely on runoff from melting snow and glaciers this will lead to reduced availability of water when climate change is likely to increase the occurrence and intensity of extreme weather events. Storms, heavy rainfall, heat waves and glacier melt will amplify hazards in mountain areas worldwide, while the melting of glaciers and the upward movement of permafrost will release loose rock and soil, aggravating the risks of rock falls, debris and mud flows and glacial lake outburst floods. Prolonged periods of higher temperatures will increase the incidence of droughts and fires, leaving some regions prone to desertification.

Warmer climates have already shifted vegetation belt and the snow line towards higher elevation. It has led to serious impact on flora and fauna, increasing the number of species that can survive at higher altitudes. As a result, rare and fragile species and those adapted to the cold may become extinct through competition and habitat loss. Extreme weather events, droughts, fires and the incidence of insect-borne diseases are all likely to increase, further threatening the habitats of mountain organisms. In some mountain regions, climate change may bring regional and local benefits. Higher temperatures could allow farmers to grow crops at higher altitudes and allow plants to produce higher yields provided that water and soil conditions are adequate. An extended growing season and accelerated soil decomposition may lead to improved nutrient intake by trees and other plants, which may in turn increase growth and productivity. However, for many mountain areas in the South, current models predict that water availability will decline and rainfall will become more erratic. Climate change will worsen the living conditions of most mountain dwellers and will also have heavy impacts on the lives of the downstream people. Crop failure and the loss of livestock will threaten the livelihoods of mountain farmers who are already vulnerable and food insecurity, while changes in snow patterns will affect the tourism industry and cause enormous economic losses. The likely increase in the number and dimension of disasters would require expensive measures to protect the infrastructure and settlements. Finally, the expected changes in water availability will seriously affect upstream–downstream relationships and are likely to lead to conflicts.

Approaches to Sustainable Mountain Development

Sustainable mountain development requires that mountain ecosystems be managed in ways that allow them to provide goods and services for local livelihoods and lowland people both in present and future times. By addressing environmental, economic, social, cultural and political issues in a holistic manner, sustainable mountain development aims at improving the lives of mountain people and the life-support systems of the surrounding lowlands. Mountain systems are complex, so any activity in mountain areas will have a number of environmental and socio-economic consequences. Different subsystems are closely interlinked; hence, management tools that tackle only a single component or segment will not be effective.

To respond to the global challenges/threats, holistic, participatory and integrated approaches that address all aspects of sustainability are required. The specific needs and inter-linkages of different aspects of sustainable mountain development, such as water, biodiversity, tourism and infrastructure, must be taken into account. To achieve sustainable mountain development, it is essential that all the concerned stakeholders are involved and that awareness is raised about mountain ecosystems, their fragility and prevalent problems, and about ways of addressing them. To ensure that all relevant ideas, experiences and contributions are considered in the search for sustainable solutions, participation should extend from the international or national to the local level, and should involve all stakeholder groups, including government officers, scientists, technicians, local communities, the private sector and non-governmental organizations (NGOs). Local livelihoods play an important role, and inhabitants of mountain regions should be actively and continuously involved in the planning, implementation and follow-up of development activities. Sustainable mountain development should take into account the experiences of indigenous mountain communities and should support traditional practices and land use systems. New technologies and approaches such as conservation agriculture and soil and water conservation techniques need to complement and be integrated into local practices. To divert the pressure on land resources and improve the livelihoods of mountain inhabitants, alternative income generating activities such as the production of high-quality products or eco-tourism should be promoted. The benefits of sustainable mountain development extend far beyond mountain regions. For example, well-managed mountain forests and the afforestation of degraded areas provide protection against natural hazards and help to regulate water flow and improve the quality of water—both of which are very important for lowland people. Yet, in general, the costs for implementing sustainable mountain development are borne by mountain communities, which create an economic imbalance. Downstream users have to be made aware of these mechanisms and provide compensatory payments to mountain communities. These communities have to be compensated for the conservation efforts and the rational use of resources that ensure the provision of environmental goods and services to lowland areas. All stakeholders involved need to reach consensus on how to value and apportion the costs of sustainable mountain development. It requires a clear understanding and appreciation of upstream–downstream linkages and tools for the equitable sharing of costs. Harsh climates and difficult, often inaccessible terrain combined with political and social marginalisation make mountain people vulnerable to food shortages. FAO estimates that 78 % of the world's mountain area is unsuitable or only marginally suitable for growing crops. In 2002, an FAO study found that 90 % of the world's mountain people (nearly 325 million) were living in developing countries or countries in transition, and that 245 million of these people (more than 75 %) were experiencing or were at risk of hunger. This figure is now likely to be even higher, given the global economic situation, soaring food prices and population growth. Nutrition studies indicate that mountain populations suffer from high rates of micronutrient deficiencies. For example, inhabitants of the Andes, the Himalayas, and mountain ranges in China are considered to be at the highest risk of iodine deficiencies. Data from the Andes and the Himalayas indicate a high prevalence of vitamin A deficiency. Hunger and micronutrient deficiencies are contributing factors for the significantly higher infant mortality rates in mountain regions. As well as being a symptom of poverty in mountain communities, hunger and micronutrient deficiencies also perpetuate poverty by reducing people's ability to work and make a living.

The impacts of climate change, environmental degradation and deforestation are increasing the hardship of mountain communities, particularly for women, who now have to travel greater distances to collect fuel and fodder; crop failure is becoming more frequent. The reported consequences include food deficits, growing outmigration and even the trafficking of mountain girls/women to lowland cities. Sustainable mountain development can be implemented at different geographical scales: from very small areas that cover only a few selected villages, to larger regions such as entire catchments, including those that cross international boundaries. Because of the fragility of mountain ecosystems, development activities should always be preceded by a careful assessment of local conditions and accompanied by impact monitoring. This is especially important when successful projects are scaled up to larger areas or applied in new settings with different characteristics.

Natural Resources Management

The scientific and sustainable management of water, soils, pastures and forests is essential for avoiding environmental degradation in mountain areas. Integrated and diversified land use systems have to be maintained and enhanced harsh climates, marked topography and diversified geological and hydrological conditions make mountain ecosystems particularly vulnerable to inappropriate natural resource management practices and environmental degradation. Erosion rates are higher and the loss of fertility through the leaching of nutrients more accentuated than elsewhere. Owing to the low temperatures prevalent at higher altitudes, plant growth and soil formation are slower, and vegetation cover is significantly less than in lowland areas. As it is very difficult to reverse environmental degradation in mountain regions, timely action is required to prevent such processes and trends through long-term approaches that combine the management of water, soils, pasture and forests. All the stakeholders involved—mountain people, the private sector, politicians and other decision makers are responsible for ensuring the use of natural resources wisely and considering the particular characteristics of upland ecosystems. Mountains' decisive role in the global water cycle means that the management of water resources requires special attention. Water management needs to be adapted to different climatic zones: where water is abundant, any excess must be drained to avoid soil saturation and water logging on sloping lands, while in semi-arid and arid areas; structures such as micro-basins are used to harvest water.

Soil and Water Conservation in Mountains

Soil and water conservation is defined as activities at the local level that maintain or enhance the productive capacity of the land in areas affected by or prone to degradation. Soil and water conservation includes prevention or reduction of soil erosion, compaction and salinity; conservation or drainage of soil water; maintenance or improvement of soil fertility, vegetation cover and quality, etc. In many regions, sophisticated irrigation systems have been developed to overcome distance and difficult terrain between the water source and the point where the water is being used. Water management must also consider upstream–downstream processes and interests. As water becomes scarce resource in many parts of the world, equitable distribution systems and improved water-use efficiency are indispensable. The reduction of erosion and nutrient loss is an important priority in mountain areas. To manage mountain soils sustainably, conservation of agriculture soil and water conservation techniques need to be encouraged. Adequate ground cover and trees with dense foliage and extended root systems offer good protection. Other measures for preventing degradation and increasing the fertility of mountain soils include reducing slope lengths and angles through different forms of terracing; controlled grazing; diversified cropping systems; intercropping; zero-tillage and the planting of nitrogen-fixing plants along contour lines. Diversified land use systems protect natural resources, maintain or even increase agro-biodiversity, and make mountain farmers more resilient to climate variability or economic crises. Many forest management approaches and techniques used in the lowlands are unsuitable for mountain forests. For example, the clear cutting of large areas or the construction of access roads can destabilise the entire mountain slopes and lead to high soil erosion. Mountain forests should be managed with an ecosystemic approach, taking into accounts the biological characteristics and different ecological functions of a forest. Diverse forest stands with a variety of species and a differentiated age structure should be maintained, and selective harvesting techniques should be applied. Such stands are much more resilient and therefore better able to fulfil their slope stabilisation and soil protection functions. The sustainable management of natural resources in mountain areas requires significant effort and investment and imposes a heavy workload on the local people. Mountain people should therefore participate in the search for sustainable solutions, to ensure that they will be economically viable and socially acceptable. Indigenous communities' knowledge and experience have to be fully considered/recognised and the causes/impacts of unsustainable natural resource management have to be carefully assessed. The ecological consequences of unsustainable natural resource management and land use practices often have underlying causes of social, political or economic nature.

Conservation Landscape

In the last 40 years, protected areas have increased between six to eightfold. Many are located in the mountain areas, and are often inhabited/used by the local people. Innovative approaches are needed to reconcile biodiversity conservation with development—two often conflicting goals—and local people must be engaged in the stewardship of their natural heritage. Conservation landscapes provide an opportunity for such an approach, as biodiversity sanctuaries are included within a pattern of agricultural and other land uses. The potential for maintaining high levels of biodiversity in combination with intensive but diversified small-scale farming is particularly valuable in areas in which high population density inhibits the establishment or extension of protected areas.

Economic Opportunities

Mountain regions face significant challenges when they interact with the market forces. Because of environmental/climatic constraints, difficult terrain and high production costs, economic activities in mountains rarely achieve the scale and profitability at par with that of the lowlands. The amount of time needed and the costs of transportation pose major barriers to trade and hamper the export of perishable goods from mountain regions, while the isolated nature of mountain communities leaves them cut off from information about markets and innovations. In mountain areas, the export of unprocessed goods at relatively low prices, such as timber or mining products, is frequently coupled with the import of expensive commodities that cannot be produced locally. In addition, mountain communities provide lowland areas with important environmental services—often without adequate financial compensation. An economic imbalance is created in which lowland societies gain most while mountain communities bear the environmental and social costs. Most mountain households pursue several activities to generate income. This diversification of livelihoods makes socio-economic systems more adaptive and responsive to the fragility of mountain ecosystems. Such diversification also can enhance mountain communities' resilience and capacity to cope with global environmental changes. Economic development in mountains should build on these diversified systems and should not promote reliance on a single economic activity. Certain high-quality niche products offer comparative advantages for economic development in mountain areas. In addition to traditional handicrafts, special products can range from farm products, such as fruits, vegetables, off-season products, cheeses and wool, to non-wood forest products, such as medicinal plants, herbs and native spices. To maximise returns, mountain communities need support in moving from the selling of raw materials to the processing and marketing of goods. Processing also increases the products' durability and facilitates their transport and sale in the lowland markets. Local micro-enterprises, suitable financial institutions and efficient marketing systems are essential for the development of mountain economies.

Mountain Environmental Services

Environmental services provided by mountain areas, such as high-quality freshwater, hydropower and disaster prevention, are often perceived as public goods, and their value is rarely expressed in monetary terms. This leads to an economic imbalance between downstream beneficiaries and upstream providers of the services. Financial mechanisms to compensate mountain communities for providing such services have to be developed and implemented. In much of the industrialised world, incentives, subsidies or direct transfer payments have become the norm, and where such financing mechanisms are not available, such as in many developing countries, payments for environmental services offer a promising alternative, particularly for the provision of drinking water. Current developments and the movement towards a green economy will increase the demand for environmental services. As long as adequate financial tools and policy instruments are put in place, this will open up new opportunities for mountain communities.

The Future of Sustainable Mountain Development

There is evidence that water, energy and food will become scarce resources in coming decades. Mountains play an increasingly important role in the provision of freshwater for a growing global population, and of energy and biodiversity conservation. Without sustainable mountain development, the world will not be able to solve the global problem of increasing water scarcity. The concept of sustainable mountain development therefore needs an important place on the international agenda. In 1992, the adoption of chapter 13 of Agenda 21 'Managing Fragile Ecosystems: Sustainable Mountain Development' at the United Nations Conference on Environment and Development (UNCED) was a milestone in the history of the mountain development. Two decades later, awareness about the global importance of mountain ecosystems, the fragility of their resources and the difficult living conditions of many mountain people has increased significantly. Nevertheless, sustainable mountain development does not yet receive sufficient attention on the international agenda, and there are still significant constraints to alleviating poverty, mitigating environmental degradation and attaining sustainable development in mountain regions. In 2000, the international community committed itself to achieving the eight Millennium Development Goals by 2015.

The Rio+20 initiatives provided an excellent platform to emphasise the global importance of sustainable mountain development, enhance awareness and increase investments. The United Nations Convention on Biological Diversity (UNCBD) Programme of Work on Mountain Biological Diversity, the United Nations Framework Convention on Climate Change (UNFCCC) and the United Nations Convention to Combat Desertification (UNCCD) also provide important mechanisms for promoting sustainable mountain development. Sustainable mountain development must be given a prominent place on the global development agenda.

Despite their importance, mountains have received little attention in global discussions of environmental and development issues. This changed in 1992 with the adoption of chapter 13 of Agenda 21 at the UNCED in Rio de Janeiro, Brazil. Chapter 13 promotes the sustainable development of mountain regions, points out the need for better understanding of the ecology of mountain ecosystems and clearly acknowledges mountains' importance for humankind. For the first time, sustainable mountain development was placed on a similar footing with other major global issues.

Chapter 13 of the Johannesburg Plan of Implementation and the Millennium Development Goals presents a comprehensive draft for the overall policy framework for sustainable mountain development at the regional and the global levels. Mountain issues are addressed in chapter 24 of the Millennium Ecosystem Assessment (2005) and in the UNCBD Programme of Work on Mountain Biological Diversity. Numerous global and regional networks have been set up, and international and regional conferences and workshops have been organised. Compared to the previous two decades mountain issues have clearly gained momentum on the global agenda. Action is needed to translate this into relevant national/regional/local programmes and existing institutions/policy frameworks should offer important platforms for promoting/intensifying sustainable mountain development.

Mountain systems need support to meet the new challenges and to benefit from emerging opportunities such as the green economy. Innovative financial mechanisms for the valuation of mountain goods and services have to be developed. Closer financial linkages between upstream and downstream areas as well as among the private sector, government institutions and local entrepreneurs are also required. The local potential for innovations should be explored and promising initiatives should be supported. There is a clear need for stronger enabling environment with more supportive laws, policies and institutions. Countries with mountain systems should join forces to make their voices heard. Food security, poverty alleviation and overall development are critically linked to mountain ecosystems and the processes in mountain regions. Sustainable mountain development is therefore essential for the well-being of not only mountain people but also of all humankind.

Reference

Millennium Ecosystem Assessment. (2005). Living beyond out means, natural assets and human well-Being, the board of millennium assessment.

Chapter 10
Conclusions: Major Problems and Prospects of Sustainable Livelihoods

Highlands and Mountains are vital economic and ecological resources, high in both biodiversity and minerals. Although they occupy only one-fifth of the earth's land surface (Pratt and Shilling 2002) and house only a tenth of its human population, more than half of the world's population is reliant on them (Earth Summit—Agenda 21, 1992), in particular for water resources and agriculture biodiversity for food production. The world's poor lives in mountain regions, particularly in the developing countries. Mountain landforms trigger precipitation, which, coupled with the water-storage capacity of glaciers, give them a vital hydrological role. Three billion people in Pakistan, India, Nepal, China and the Indochinese peninsula depend on rivers that flow from the Himalayas. Similarly, many millions rely on the headwaters of the Amazon in the Andes, and on the Nile and Congo rivers, which flow from the East African highlands. In addition to water, 38 % of the world's forest is located in the mountain areas. Mountains are the prime source of most minerals; they house a large portion of the earth's biodiversity, representing some 40 % of the critical biodiversity surface area (Pratt and Shilling 2002). Mountains are also home to many indigenous peoples with their valuable cultures and indigenous knowledge. Notwithstanding the enormous importance of these regions, many highland and mountain communities remain isolated, trapped in poverty and live at the margins of society. Mountain people are generally isolated and depend on highly diverse indigenous cultures and knowledge for their livelihood and management of natural resources.

After years of negligence, sustainable agriculture is again emerging as a central theme in rural development. Given the importance of mountain development for the people who dwell there, as well as the importance of mountain resources to the world as a whole, the renewed attention brings new optimism and opportunities. However, the concept of development this time must go beyond the concern with supply, but be also concerned about the participatory approaches to understanding problems and opportunities facing farmers and farming communities. It must balance production with stewardship of resources and improved income-generation opportunities. Sustainable agriculture, therefore, encompasses the economic, bio-physical, socio-cultural and environmental aspects of development.

V. P. Sati, *Towards Sustainable Livelihoods and Ecosystems in Mountain Regions*, Environmental Science and Engineering, DOI: 10.1007/978-3-319-03533-8_10,

A significant proportion of mountain people live on less favoured lands, such as in the upper watersheds of the Andes and the Himalayas, and in the East African Highlands, where severe degradation of resources is experienced (Hazell and Garrett 1996). Living far away from the centres of commerce and power, mountain people of the developing world exert little influence over the policies and decisions that influence their lives. Mountain nations such as Tibet, Lesotho, Nepal, Rwanda, Burundi and Ethiopia are among the 20 poorest in the world. In these regions, the majority of the population is under 18, and literacy varies widely, but is generally very low. Household income is also low—ranging from $150 to $500 per capita per year. Shocks, resulting from natural disasters or political actions, disease, war and market forces have profound impacts on the livelihoods of mountain people (Kherallah et al. 2000), and food insufficiency can vary from 1 week to more than 3 months per year. The areas where consecutive shocks are experienced in the form of droughts, floods or pests, vulnerability is increasing. Given that such events are frequent, food insecurity is a major problem. As a result of this the infant mortality ranges from 15 to 71 per thousand, and life expectancy is low, particularly for women.

Local government and communal institutions exist in many areas to manage natural assets, such as water and land, but their access to higher echelons of government is very limited, and their voice in policy is often not heard. This is because of their remoteness, and the history of neglect of these societies (Pratt and Shilling 2002). There is a great diversity in the characteristics of the populations of mountain peoples of the developing world. Population growth rates and density of population vary from one mountain region to another. For example, population density in the Himalayan region is low while population growth rate is high.

Mountains are among the harshest of environments to live in, and often marginalised people such as indigenous communities and cultural minorities live in these difficult terrains. Making a living there is challenging and increasingly men are migrating to cities in search of jobs, leaving women, children and the elderly to maintain the homestead. Women share up to 50 % of the workload in crop production and farm management, and up to 100 % in livestock production responsibilities, in addition to other traditional household roles. As a consequence, mountain poverty has become increasingly the plight of women. Despite their additional responsibilities and dependence on farming income, few women are given title to land and fewer still, access to financial credit. In some societies women's movements are restricted with deleterious consequences for education and extension opportunities.

Ecosystems are the basis for social and economic development. Although the interaction and interdependency between the dynamics of human and ecological systems is well recognised, the impact of settlements and economic activities on the supply of goods and services that mountains provide to society has been less studied. It has been wrongly assumed that responses of ecosystems to human use are linear and predictable, can be controlled, human and natural systems are separate and can be treated independently (Folke et al. 2002). Logically, these natural and human systems behave in nonlinear ways, exhibit marked thresholds in their dynamics, act as strongly coupled integrated systems and present spatial and temporal variability inherently scale-limited.

An important challenge for sustainable mountain development is the eradication of political marginalization. There is need to develop and enforce mountain-specific legislation and policies that allow mountain people's participation in decision-making, the integration of upstream and downstream interests and the consideration of transboundary issues. Most mountain regions share long histories of political marginalisation. Mountain people have limited access to policy and decision-making beyond the local or district levels, and their development is often externally driven. So far, only a few countries have adopted mountain-specific policies or laws. Instead, mountain issues are commonly addressed through sectoral legislation related to agriculture, forestry, land or water. Such policies and laws are normally developed and negotiated with a lowland focus and pay inadequate attention to the fragility and specificities of mountain ecosystems and the needs and interests of mountain people. Decision-makers based at the lowland centres tend to perceive mountains as hinterlands that supply resources, and fail to consider the specific challenges of mountain development. To achieve political, economic and social integration, mountain people's voices have to be heard and mountain communities have to be recognised as equal partners in policy and decision making.

Decentralisation is one of the best ways of empowering mountain inhabitants, reducing costs and increasing efficiency, but local administrations often lack the necessary resources and expertise to fulfil their new functions, especially in the developing countries. In many places it is necessary to establish new institutions, strengthen existing ones and enhance the capacity of local stakeholders. Sufficient monetary resources must also be transferred from the central to the local level. The sustainable development and protection of mountain regions and the improvement of local livelihoods should be at the core of mountain legislation. Policies and laws must take into account the specific characteristics, needs and challenges of mountain areas, while also considering the broader political context, including downstream interests. Key aspects to be addressed are the improvement of infrastructure and the enhancement of public services, particularly in the health and education sectors. To meet such objectives, legislation has to provide adequate financial instruments and resources.

Agriculture of mountain region is inter-linked with farming, animal husbandry and natural resources (Rao and Saxena 1994; Maikhuri et al. 1996; Sati 2004a, b, c, d). However, chemical analyses of all these crops prove as highly nutritive food value and safe diet (Rawat and Bhatt 2002). The production and consumption of these crops will be promoted for better health and nutritional security. It is true that food habits change with the time and economy of a region. The Himalayan region is known as a repository of varieties of crop species. In recent years the crop diversity of the region has declined to alarming proportions. Many factors are responsible for the sudden decline of cropping pattern, viz. change in food habits, loss of biodiversity and natural resources, low grain productivity and replacement of traditional cropping system (Semwal et al. 2001).

Legislations need to address the protection of ethnic minorities and the cultural heritage of mountain people, and to recognise community-based property rights. In many countries, failure to recognise the traditional land-use rights of mountain communities has opened the way for inequitable extraction and use of natural resources and has led to disputes. With secure land titles and access to local resources, mountain people have greater interest in managing them sustainably, and are in a better position to negotiate compensation payments, such as for the extraction of timber or mining products, or the establishment of PES schemes. Mountains have significant life-support systems, aesthetic/cultural values and economics of the people residing in or adjacent to the area. They have depended directly or indirectly on mountain resources for livelihoods from time immemorial. With the rapid increase of industries and other human pressures in the plains, mountainous regions have considerably damaged the ecology and natural resources of the area. Mountainous regions have ecological potential with natural resources but they are economically less developed. These resources need to be utilised in a rational and sustainable manner for long-term sustenance for the people of this region.

Women play a vital part in mountainous regions. Women are very hard-working and they toil through the day, starting with the family works, nurturing children and livestock, going out for fodder, fuel, drinking water and NTFPs collection for sustaining livelihoods. Unfortunately, these women do not possess the status of farmer. If women are given the status of farmer, they will get rights over their land, which will go a long way to empowering them and improving their status. In the midst of all these worries, a hill woman remains firmly devoted to her work, keeps her composure and is honest. Enhancing social security for these women and setting up processing industry and management of natural resources, medicinal plants cultivation, etc., can enhance employment opportunities both for men and women in this region, which will result in less migration of the menfolk. Remote areas have inadequate networking of transport and other basic supporting facilities, which come in the way of proper benefits of agriculture, horticulture and other developmental programmes conducted by the state and other organisations in the ICHR (Uniyal 2006).

Mountain Specificities

Constraints

In terms of constraints, mountains are characterised by inaccessibility, fragility and marginality. It should be noted that these specificities are not only interrelated in several ways, but the locations in mountain areas are equally inaccessible, fragile or marginal. Nevertheless, they help to shed light on some of the key constraints facing the development of mountain areas.

Inaccessibility: Its concrete manifestations are isolation, distance, poor communication and limited mobility. Besides the dominant physical dimension, it has socio-cultural and economic dimensions, which are reflected by socio-economic differentiation and inequity of access to resources, information and opportunities. Inaccessibility exaggerates other conditions such as marginality and diversity.

Fragility: Due to altitude, steep slopes and other biophysical conditions, mountain areas are vulnerable to degradation. In most cases, the damage is irreversible or reversible only over a long period.

Marginality: Mountain areas tend to be marginal in terms of their remoteness and physical isolation, fragile and low-productivity resources. There are also several man-made handicaps, which prevent participation in the mainstream pattern of activities. Mountain communities have been at the periphery of decision-making in the context of overall development in many countries. Development policies and programmes have consistently failed to give careful attention to mountain areas. Even when attention has been given, approaches have been inappropriately organized and managed, resulting in many adverse impacts—particularly on the environment. Marginality is also evident in the influence of non-local factors in the use of mountain resources. Today, international markets determine how local resources are used in many mountain areas. With weak institutions and inadequate skill development in mountain areas, there is a great danger that this will continue in the future.

Potentialities

Despite the constraints characterising mountain areas, the diverse nature of the culture, resource base/environmental conditions creates potentials for products and activities that have a comparative advantage. It offers a number of opportunities for resource and product-centered activities, which could enhance both productivity and human welfare on a sustained basis. Mountain agriculture provides indeed much more than the mere production of food, fibre and commodities. When practiced in a sustainable way, it assures food security, environmental stability, conservation and sustainable use of the biological resources. It also ensures rural income/employment, rural–urban population balance, social equity, cohesion and maintenance of cultural features/identity in the overall context of globalisation. Mountains provide a wide range of goods and services essential for guaranteeing the availability of vital resources to the rest of the society in relation to water, biodiversity, landscape, risk prevention, specific products and culture, resulting in a series of interdependencies between upland and lowland areas. There usually tends to be a lack of recognition of the importance of these inter-linkages in the institutional and policy frameworks of countries. Thus, the promotion of sustainable mountain development can play an important role to benefit lowland areas by ensuring adequate supplies of water, environmental stability, conservation of biodiversity, rural–urban population balance, etc.

Diversity

Diversity is an innate feature of mountain areas as one can find immense variations among and within eco-zones, even within short distances. Not only is there diversity in terms of the degree to which inaccessibility, fragility and marginality affect different mountain areas; more importantly, mountain areas are home to a large number of biological resources. They provide an important source for water and unique products such as fruits, spices, flowers, medicinal plants, minerals, etc. There is also diversity in terms of the economic, social, cultural and institutional set-up of mountain areas. The socio-cultural diversity leads to variations in social organisation, attitudes and methods of managing collective goods and community resources. The harnessing of this diversity in a sustainable manner is therefore an important issue for mountain areas.

Mountains have valuable medicinal plant diversity because of the majestic Himalayan range and a healthy ecosystem. So far 1,748 medicinal plant species have been documented from the Indian Himalayas (Samant et al. 1998). On the regional scale, the maximum medicinal plants have been documented from Uttarakhand, followed by Sikkim and North Bengal (Kala et al. 2004). A majority of the medicinal plant species are endemic to the Himalayan region from all documented species of different national bio-geographical zones. The rural community of Uttarakhand generally uses some natural resources as plant species as a source of food, fodder, timber, medicine and various ethnobotanical purposes and source of income generation activity (Uniyal and Shiva 2005). Medicinal plants have also strong acceptance in religious activities in Uttarakhand; the locals worship plants in the form of various gods and goddess. Some of the plant species, viz. *Saussurea obvallata, Ocimum sanctum, Cedus deodara, Juniperus communis, Pinus roxburghii, Musa paradissica, Zanthoxylum armatum, Ficus benghalensis* and *Ficus religiosa* are examples of medicinal plants highly used for medicinal as well as religious purposes in this region (Silori and Badola 2000; Kala et al. 2006). Medicinal plants are an integral part of the life of the local communities of Uttarakhand. Local people and traditional healers have valuable information about indigenous knowledge of their healing capacities. With diverse climatic condition of state, the region is suitable for cultivation of medicinal plants. In Uttarakhand, about 500 medicinal plant species are being traded. Nearly 160 of these have been identified as threatened species; most of them are found in the high altitude areas of this region (Nautiyal and Nautiyal 2004). In this region, cultivation of medicinal plants has been initiated by various institutions and organizations, therefore, it has been declared as herbal state. Some medicinal plants grow in the wild in very specialized micro-habitats and association. Only selected plant species are possible for cultivation in agriculture fields for commercial and alternate income generation options.

Major Constraints to Sustainable Livelihood

Lacking in Infrastructural Facilities

Lacking in infrastructural facilities such as cold storages and proper marketing system is a major constraint that the products do not receive full returns. Often they are unsold and consumed locally. Establishment of cold storages according to the production and distance between the producing areas will promote the production of cash crops and will motivate producers to cultivate them. Furthermore, it will also ensure proper marketing of products. Many producers of cash crops in Janglechatti village have requested the government to construct a cold storage. Similarly, small-scale industries and food processing centres should be established and local people involvement should be ensured.

Lacking in Holistic Government Initiatives

Institution involvement in terms of encouraging farmers to cultivate certain cash crops and providing them financial assistance, insurance of their crops at the time of crop failure and prepaid cash of their crops are lagging behind in the region. HAPPRC of the HNBGU initiated such practice in a very high altitude village of Ghais where farmers are growing medicinal plants and HAPPRC works as mediator between the growers of medicinal plants and the companies. Government initiatives for any development programmes have multiple impacts on the area, particularly at the time when the government has to control the entire affairs of the state. The impacts may be negative or positive depending on the degree of involvement and rationale of planning. Political instability (untimely change in government) and lack of involvement towards development planning further limits the choices and options of livelihoods. Due to its remoteness, very less impact of policies towards development could reach in the region and consequently, it remained untouched by the waves of modernisation and industrialisation.

Lack of Proper Marketing

Lack of proper marketing of various cash generating products in the region has remained a major hurdle to growers. Products of high quality and quantity do not get a proper market for a long time and consequently, growers do not get the actual returns. Infrastructural facilities as transportation and cold storages are slackened. There are instances when crops after taking a year to get a form of product are not sold out. Discussions with growers of medicinal plants, fruits and vegetables regarding their products and economic viability has revealed that growers do not find them easy to grow; instead they prefer cultivating traditional cereal crops. Villages where agro-climatic conditions are suitable for growing cash crops, and where farmers intend to grow them extensively, are highly inaccessible. A good marketing system (from regional to national) will be the base for exporting these products. The upland areas are lagging behind in a good marketing system, therefore the products do not get sold in a proper time bounded manner. The intention of growing subsistence cereal crops is increasing as these crops are the main source of livelihoods and they are self-reliant and do not require services from outsiders.

Lack of Training Programmes

Diversities in programmes involved in enhancing and diversifying livelihood options can be noticed in this region. Some of them are decades old. In the 1970s, the then Government of Uttar Pradesh launched a programme that aimed to demarcate the boundaries of middle and high hills of UP (now Uttarakhand) as *fruit belts* keeping the suitability of agro-climatic conditions in view. This programme was also initiated due to the impressive progress achieved by Jammu and Kashmir and Himachal Pradesh. However, this scheme was not successful as the Department of Forest did not support it. Meanwhile, in some areas of middle and high altitudes, the local farmers initiated cultivating fruits. Transformation of cereal cropped land into fruit crops took place. Almost for a decade, apple and citrus dominated the cropping pattern in these areas. Now the situation has changed. The land which was under fruit crops are either abandoned or replaced by other crops. This was due to improper nourishment of fruit trees. It is important to note that the State Government and the Department of Horticulture appointed master trainers to impart training to growers, but the trainers do not perform their duties. Farmers do not know technical aspects because they are not well trained, therefore the situation becomes sternly serious.

Limited Understanding of Mountain Environments

Current understanding of socio-economic, institutional and biodiversity processes in mountain areas is still very limited. The sporadic bursts of attention given to mountain areas have generally been a post-disaster phenomenon. Systematic attention to these problems is a very recent phenomenon, and it is extremely limited in extent and coverage. There are large gaps in understanding issues specific to mountain environments. The persistent knowledge gaps in understanding and delivery of sustainable mountain development were identified as an area of major concern by UNCED Agenda 21, Chapter 13.

Institutional Gaps

The limited development efforts that have taken place in mountain areas often had little impact in terms of poverty reduction and led to worse environmental conditions. There has been failure in mobilizing the local community organizations, and the expansion of central development bureaucracies is limited and their performance is poor. Stronger local-level development organisations are indispensable for successful management of development activities in view of the various constraints of access and communications. Decentralised and participatory organizational frameworks have been accepted as needed, but efforts to promote these need to be sustained. Organisations working to develop this aspect in the fields of education, functional literacy, skill improvement, female access to educational and training opportunities, etc., are still very limited. The real challenge is to ensure that capabilities and capacities at the community and other levels grow on a sustainable basis, that economic changes do not result in cultural deprivation and that technology gains do not result in environmental disaster.

Climate Change

A study of the climate change phenomenon, particularly in the mountain regions, requires an appraisal of climate data of the past one or more centuries. Currently, the perception of all stakeholders towards the impact of climate change in mountain regions is unanimous. As discussed in the previous paragraphs, the land which was used largely for fruit cultivation is currently abandoned. The grower's perceptions are that the unproductiveness of fruits plants is due to climate change. The areas where intensive cultivation of apple was carried out during the past are no more for its cultivation. This belt has been shifted greatly to higher elevations. This is also the case with citrus, nut-stone fruits and other agricultural crops.

During my extensive field visits of the Gwaldom-Lolti apple fruit belts, I observed that the apple orchards were completely destroyed. This area was producing large quantity and quality of apple during the 1980s and, apple was sold in the national market. This occurred mainly due to increase in the temperature in the lower altitudes. A similar situation was noticed in the distributional pattern of pine and oak forests. Pine trees grow in warm climates whereas oak trees grow in temperate climates. Due to continuous changes in climatic conditions and as the temperature is increasing in the valleys and mid-altitudes, the pine forests have been shifted towards higher elevation and they have already invaded the areas where oak forests were grown. This situation has led to disappearance of oak forests from their original realms. Disappearance of oak forest was also due to its overexploitation because oak trees have multiple uses from being used as animal fodder to fuel wood for cooking food, and warming the houses during the winter season. It has also an impact on the cropping pattern.

There are bitter experiences of failure of individual efforts towards diversifying and enhancing livelihood options. The agricultural land is largely fragmented. If innovative ideas are implemented by farmers for changing the farming systems they are not yielding any results. Under such circumstances, what are the steps needed for enhancing and diversifying the options for better livelihood? There is a general consensus among producers, researchers and academicians about community participation. Self-help groups and community participation in development activities should be ensured. Although there are various schemes launched for assisting farmers and organising community groups for development purposes, a major thrust is required to assist the community groups as a whole and not as individual farmers. If a group of community people participate together in any development activities, the entire scenario will change and development of the region will take place.

Land abandonment can be observed in many locations. It is generally found in the villages located in the lower altitudes or along the roads. One of the most important factors of land abandonment along the road is commercial uses of land or for constructing settlements. The region is best known for its tourism, especially for pilgrimage tourism. Farmlands along the roads have been converted into accommodation avenues; lodging and boarding. Along the valley of Alaknanda River, from Devprayag to Karnprayag or in other river valleys, around 80 % land has been abandoned and mushrooming of hotels, motels, lodges and *dhawas* can be seen. In villages at lower altitudes, land abandonment is due to large-scale migration of the populace to the urban centres of India. They are permanent migrants and settle in the urban centres. It was also noticed that people who earn money as remittances also leave their agricultural land abandoned. A common trend was observed in the villages where only those households that are fully dependent on the output from the agricultural land to carry their livelihoods, cultivate cereals and also plough the agricultural lands of other people who have migrated.

Opportunities and Options for Sustainable Livelihoods

Mountain regions have immense potential for developing land-based resources especially for the extensive cultivation of agricultural and horticultural crops. Agro-climatic conditions such as temperature, rainfall, humidity, soil fertility, water and landscape are suitable for cultivating agricultural and horticultural crops.

Agricultural Diversification

Agricultural diversification has many impacts as follows, viz. first it sustains soil fertility and second it enhances livelihood options. Diversification in agriculture has already been in practice in highland areas. Villages located in the highlands (above 1600 m) have the capacity to store grains for adverse times. Even if the market collapses, these villages are quite able to sustain their livelihood. Global changes can be noticed everywhere in mountain regions. Population has increased manifold. Changes in farming systems can be observed. Cereal crops farming was not enough to meet the increasing food demands resulting in changes in cultivation from millets to paddy and wheat. During the past, paddy and wheat crops were rarely sown but now it is very common food. I personally do not agree with Rao and Saxena's (1993) statement who state, 'food insecurity was due to change in food habits of paddy and wheat instead of cultivating millets'. During conducting this study in the field, I experienced that the cultivation of wheat and paddy has increased the production pattern and obtained food security to a certain extent as the people have become self-reliant after they changed their land from the cultivation of millets to the cultivation of paddy and wheat. Even today, a large proportion of sown area is devoted to the cultivation of paddy and wheat crops in the entire ICHR. During the 1980s and 1990s, large-scale cultivation of fruits and off-season vegetables was carried out. The production of potato and onion was increased during this span of time but, later on, the production was decreased and at present the maximum cultivable land is devoted to the cultivation of paddy and wheat. The increasing trend of cultivating paddy and wheat crop shows that these crops are good enough to maintain food security as they are grown traditionally and the crops are mostly rain-fed.

Use of Non-timber Forest Products

NTFPs such as bee keeping, herbs and medicinal plants, oak bark, natural dyes, resin, wild fruits, firewood and fodder are the major options for livelihoods enhancement in the mountainous regions of the world, especially in the ICHR. If sustainable use of NTFPs is made, it will have trio impacts, i.e. economic development, employment augmentation and it will reduce migration to a great extent. From the discussion in the previous chapters it can be concluded that the populace of the mountainous regions is fully dependant on the outcome from agricultural practices; however, the production of cereals is not sufficient to fulfil the daily food requirement of the populace as the cultivation method is traditional and production and per ha yields of crops is considerably low. Agricultural practices are still carried out on trial-and-error method. Farmers of the region have tried to get rid of the menace of food insecurity and malnutrition and they have already preferred alternate farming systems, i.e. from the cultivation of subsistence cereals to cultivation of paddy and wheat, and during the recent past, also the cultivation of cash crops though they did not achieve success in all these practices. There are various factors that have affected these practices. These are (i) narrow and terraced fields with low soil fertility, fragile landscape that does not permit to use modern and innovative technology in the farmlands, (ii) low production and productivity of crops, (iii) lacking in market, cold storages and other infrastructure facilities, (iv) lacking in peoples' participation and (v) finally the incompetency of government officials in the process of agricultural development. In the ICHR, community participation is negligible in the developmental process and therefore, any efforts for enhancing livelihoods and agricultural development could not receive the expected outcome; whereas at the global level, community participation in development activities gained outstanding results. Harnessing the niche-based NTFPs sustainably along with biomass-based agricultural products will further enhance the livelihoods of the populace. Development of tourism and hydroelectricity may be a promising sector as the ICHR is bestowed with plenty of natural locales and huge water resources.

Alternate Occupations

Alternate non-farm livelihoods could help in absorbing the growing Himalayan population and their growing needs. Occupational diversity has found to be significant catalysts for development of the rural areas. If diversity in occupation is adopted keeping the adaptability in mind, livelihoods enhancement can be achieved. This would also have the benefit of reducing the growing pressure on natural resources and resultant ecological degradation.

Value Addition

The ICHR is among the most natural resource rich areas of the world; however, it is very far from the mainstream, and also far from technologies and markets, therefore the whole region is a totally primary produce-based local/regional economies. Although some primary produce are exported to distant markets, yet most of the produce is used domestically and sold in the local market. Value addition in the varieties of produces, which are largely unutilised, can enhance the value of the products and can enhance the income level of the marginal farmers.

Policy Interventions

Developmental interventions through government agencies need a considerable policy framework. Crop diversity needs to be conserved, but at the same time, a considerable proportion of cultivable land is to be devoted for cash crops. Both situations are favourable for sustainable livelihoods. For centuries, the local people have been practicing different systems of farming in this hilly region as trial and error. No doubt, sustainability in livelihood had been achieved to a certain extent, but the hardship, particularly of women has not reduced. In many areas, especially in the highlands where development of the infrastructural facilities is marginal, the populace is still struggling for basic needs of food, clothes and shelter that are still not easily accessible. Several studies have been carried out by researchers and academicians towards enhancing and diversifying livelihood options. The government has also initiated people supportive programmes for speedy development and welfare of the local people. Earlier, all segments of the society had perceptions that due to the mountain specificities and uniqueness of this region, development could not take shape. Further, the development policies were not specific to the specificities of the region. Even after the formulation of Uttarakhand as a separate state, the whole region has remained underdeveloped. Thirteen years have been passed and no changes have been observed in the pathways of development. Under such circumstances, policy intervention for sustainable development is the only key that will lead to overall development of the mountain regions. The following suggestions are given for speedy development:

1. Development of infrastructural facilities such as transportation, hydroelectricity generation, hotels and housing without damaging the ecology of the region.
2. Management of forest resources for the well-being of mountain people should be the priority of the forest department as well as of the local government; for this purpose people's participation should be ensured.
3. Sustainable use of natural resources and mountain niche products, which are abundantly found in the entire mountain regions of the world and are the major source of livelihoods of the mountain people, should be ascertained and natural resources should be properly managed.
4. Efficient and sustainable use of landscape, biospheres/sanctuaries for the construction of dam, development of tourism and for the use of timber/non-timber forest products/medicinal plants will ensure sustainable livelihoods and well-being of the marginal mountain people.
5. Education for better livelihoods through capacity building to absorb new technology and innovation.

References

Folke, C., Carpenter, S., Elmqvist, T., Gunderson, L., Holling, C. S., Walker, B. et al. (2002). Resilience and sustainable development: Building adaptive capacity in a world of transformations. Scientific background paper on resilience for the process of the world summit on sustainable development on behalf of the environmental advisory council to the Swedish government. p. 33.

Hazell, P. & Garrett, J. (1996). Reducing poverty and protecting the environment: The overlooked potential of less-favored lands. 2020 Vision Brief 39. Washington DC.

Kala, C. P., Farooquee, N., & Dhar, U. (2004). Prioritization of medicinal plants on the basis of available knowledge, existing practices and use value status in Uttaranchal, India. *Biodiversity and Conservation, 13*, 453–469.

Kala, C. P., Dhyani, P. P., & Sajwan, B. S. (2006). Developing medicinal plants sector in northern India: Challenges and opportunities. *Journal of Ethnobiology and Ethnomedicine, 2*, 23.

Kherallah, M., Delgado, C., Gabre–Madhin, C., Monit, N., & Johnson, M. (2000). *Agricultural market reforms in sub-Saharan Africa: A synthesis of research findings*. Washington DC: Markets and Structural Studies Division IFPRI.

Maikhuri, R. K., Rao, K. S., & Saxena, K. G. (1996). Traditional crop diversity for sustainable development of central himalayan agro-ecosystems. *International Journal of Sustainable Development and World Ecology, 3*, 8–31.

Nautiyal, M. C. & Nautiyal, B. P. (2004). Collaboration between framers, research, institutions and Industry: Experience of *Picrorhiza kurrooa* cultivation at Gheshe village in Chamoli district, Uttaranchal. *Searching Synergy*, Royal Tropical Institute Bulletin, The Netherlands, *356*, 63–71.

Pratt, D. J. & Shilling, J. D. (2002). High time for mountains. A program for sustaining mountain resources and livelihoods. World Development Report: Background Paper.

Rao, K. S., & Saxena, K. G. (1993). *Sustainable development and rehabilitation of degraded village lands in Himalaya*. Dehradun: India.

Rao, K. S., & Saxena, K. G. (1994). *Sustainable development and rehabilitation of degraded village lands in Himalaya*. Dehradun: Bishen Singh Mahendra Pal Singh.

Rawat, R. S., & Bhatt, V. K. (2002). *Nature's pharmacopoeia* (p. 148). Dehradun: Navdanya Publication.

Samant, S. S., Dhar, U., & Palni, L. M. S. (1998). *Medicinal plants of Indian Himalaya: Diversity distribution potential values*. Almora: G.B. Pant Institute of Himalayan Environment and Development.

Sati, V. P. (2004a). Systems of agricultural farming in the Uttaranchal Himalaya, India. *Journal of Mountain Science, 2*(1), 76–85. http://www.imde.ac.cn/journal

Sati, V. P. (2004b). Systems of vertical horticulture in the Alaknanda Basin of Garhwal Himalaya. *Annals of Nagi, XXIV*(2), 49–58.

Sati, V. P. (2004c). *Horticultural development in hills: A Case for the Alaknanda Basin* (Ph. D. Thesis). New Delhi: Mittal Publications. www.vedamsbooks.com/38930

Sati, V. P. (2004d). *Uttaranchal: Dilemma of plenties and scarsities*. New Delhi: Mittal Publications. www.vedamsbooks.com/38930

Semwal, R. L., Maikhuri, R. K. & Rao, K. S. (2001). Agriculture-ecology, practices and productivity. In O. P. Kandari & O. P. Gosain (Eds.) *Garhwal Himalaya, nature, culture and society*. Srinagar: Trans Media.

Silori, C. S., & Badola, R. (2000). Medicinal plants cultivation and sustainable development: A case study in buffer zone of the Nanda Devi Biosphere Reserve, Western Himalaya India. *Mountain Research and Development, 20*(272), 279.

Uniyal, B. & Shiva, V. (2005). Traditional knowledge on medicinal plants among rural women of Garhwal Himalaya. *Indian Journal of Traditional Knowledge*. National Institute of Science Communication (CSIR), New Delhi, *4*(3), 256–266.

Uniyal, B. (2006). Traditional knowledge and conservation of medicinal plants for sustainable livelihoods in the Garhwal Himalayan region of Uttaranchal, India. In *International symposium towards sustainable livelihoods and ecosystems in mountainous regions*, 7–9 March, Chang Mai, Thailand.

Illustrations Representing Abundance Natural Resources and Livelihoods

Photo 1 A panoramic view of the middle watershed of the Pindar Basin

V. P. Sati, *Towards Sustainable Livelihoods and Ecosystems in Mountain Regions*, Environmental Science and Engineering, DOI: 10.1007/978-3-319-03533-8,

Photo 2 Dense pine forest in the upper watershed of the Pindar River

Photo 3 The Great Himalayan Ranges, dense oak forest, agriculture land and settlements in the Lastar Gad sub-watershed of the Mandakini River

Photo 4 Mount Chaukhamba, panoramic view of the middle watershed of the Pindar River and wheat crop is grown in village Kaub (1,800 m)

Photo 5 A village, located at 1,900 m near Gwaldom town is surrounded by the dense pine and oak forests and the agricultural fields in a gentle slope

Photo 6 A scenic view of the terraced fields and the settlements in the Ata Garh sub-watershed of the Pindar Basin

Photo 7 The Alaknanda River is flowing near Srinagar town

Photo 8 A huge landslide zone and the settlements in its surroundings in Mandal region of Chamoli District

Photo 9 Women are carrying fodder (*green* leaves of oak) near Gairsain town

Photo 10 Water scarcity amidst of an abundance of water resource

Photo 11 Mustard seeds are grown traditionally and are important oilseeds. All the agricultural related works are done by the women and they are known as the backbone of economy

Photo 12 Women are carrying *green* leaves (*kandali*) which are used as fodder for animals and also as *green* vegetable

Photo 13 A perennial spring which is a major source of drinking water

Photo 14 Children are carrying firewood from the Raja Ji National Park

Photo 15 Keshavprayag where the Saraswati River meets with the Vishnu Ganga near Mana village, located 3 km away from the Badrinath pilgrimage. It is the first prayag amongst the seven prayags of the Garhwal Himalaya

Photo 16 The Alaknanda River meets with the Bhagirathi River at Devprayag and from here; it is called 'Ganga'

Photo 17 Millet, locally known as '*jhangora*' is one of the *barahnajas* (twelve grains) and grown extensively in the highlands

Photo 18 An agricultural field where twelve grains (*barahnaja*) are grown together and presents high agro-biodiversity in the highlands

Photo 19 Banana plants are grown extensively in the Takori Gad sub-watershed of the lower Alaknanda Basin

Photo 20 Paddy crop is now ready for harvesting in the low-lying areas of the lower Alaknanda River

Photo 21 A scenic view of the river valley terraces with paddy crops, grown in Maletha village on the right bank of the Alaknanda River

Photo 22 High yield variety calves in the Bhararisen Dairy Farm near Gairsain town in Chamoli district

Photo 23 Seeds are traditionally used for maintaining agro-biodiversity

Photo 24 An improved variety of wild fruit (*amla*) is grown extensively in the *lower* and *middle* altitudinal zones of the ICHR and is used to make Murabba, pickles, squash and is used as medicine

Photo 25 An edible wild fruit (*Timli*) is used for making pickles and vegetable curry, is grown wild between 800 and 1,500 m

Photo 26 A bunch of sweet peach is utilized domestically and it has high potential in enhancing livelihood, if it is grown commercially

Photo 27 Orange is one of the major citrus fruits, is grown between 800 and 1,600 m in all the river basins in the ICHR

Photo 28 Agricultural fields in Janglechatti village where wheat crops are grown and in the edges of agricultural field, apple is grown. Apple is mostly grown between 1,600 and 2,200 m

Photo 29 Cultivation of off-season vegetables in the Khanda Gad sub-watershed of the lower Alaknanda basin

Photo 30 A Nepali emigrant is dressing cauliflower in the Khanda Gad sub-watershed of the lower Alaknanda basin

Photo 31 Garlic, one of the spices, is grown largely in Janglechatti village located in the highland about 1,800 m

Photo 32 A seasonal vegetable, locally known as *lankuda,* is grown wild in wet areas and a major food of the rural people is sold during the monsoon season. This picture was taken from the bus stand at Karnprayag

Photo 33 A man is selling wild fruit (*kafal*) at Srinagar town. *Kafal* is an edible wild fruit, is ripened during May and June and is grown over water dividing areas of the Alaknanda River and its tributaries

Photo 34 A boy is selling strawberry at Saharanpur Chauk in Dehradun city. Strawberry is extensively grown in the Garhwal region

Photo 35 Bottle packed orange and *buransh* (a wild flower) juice is processed in the fruit processing and preserving centres with the help of self help groups at the local level and it is sold in the regional market

Photo 36 *Geranium*, a herbal plant is grown in the poly house in a bio-village near Gwaldom town in the *upper* Pindar basin

Photo 37 A newly established tea garden in Kalimati village in the *upper* water parting region of the Ramganga

Photo 38 A highland natural water pond is being used for fisheries in Gwaldom town

Photo 39 Women are being empowered slowly but steady after their involvement in the various enterprises. A woman of Khanda Gad village is running traditional improved water mill for wheat grinding and she is able to enhancing livelihoods of the family

Photo 40 A small-hydropower project generating 2 × 2.5 mw electricity at the Kail Ganga near Deval town. Small-hydropower projects are feasible in terms of landscape stability and generating electricity to enhance livelihoods

Appendix

See Tables 1, 2, 3, 4, 5, 6, 7, 8, 9, 10, 11, 12, 13, 14, 15, 16, and 17.

V. P. Sati, *Towards Sustainable Livelihoods and Ecosystems in Mountain Regions*,
Environmental Science and Engineering, DOI: 10.1007/978-3-319-03533-8,

Table 1 Land use pattern in the Alaknanda Basin (ha)

Name of DB	Total area	Forest land	Cultivable waste land	Fallow land	Other fellow	Uncultivable land	Land other than agriculture	Pasture land	Land under horticulture	Net sown area
JMT	245,877	39,774	30,532	106	292	133,132	4,068	17,849	17,842	2,282
DSL	28,504	6,213	563	43	103	7,729	435	4,838	3,338	5,242
GHT	40,997	16,794	4,717	93	193	2,878	358	6,811	3,736	5,417
KPR	13,366	5,932	1,657	67	84	701	123	2,212	802	1,788
NBR	18,750	4,143	478	38	100	1,787	611	3,959	2,732	4,902
GRS	37,672	12,867	5,638	93	329	3,529	240	7,184	2,742	5,050
TRL	15,333	6,537	993	40	84	1,068	78	1,565	1,260	3,708
DBL	24,592	9,970	2,156	146	93	3,791	129	1,923	4,514	1,870
PKR	27,150	8,225	1,210	133	139	3,933	2,004	3,356	4,032	4,118
OKM	52,778	27,577	2,527	7	122	12,499	1,371	1,654	2,897	4,124
AGM	131,419	68,668	6,293	17	305	31,122	3,415	4,119	7,212	10,268
JKL	59,511	31,095	2,850	8	138	14,094	1,546	1,865	3,266	4,649
DVP	37,398	13,506	650	1,190	1,218	477	10,566	59	11	9,721
KTN	57,704	40,039	288	541	1,110	344	6,763	30	11	8,578
KOT	41,633	24,061	1,950	575	1,262	4,266	1,440	1,323	2,059	5,213
PRI	41,584	22,320	1,912	575	1435	1,987	1,373	1,964	5,271	4,955
KRS	40,023	27,176	1,516	575	804	1,770	1,353	1,189	1,899	3,735
KPT	106,865	65,885	7,598	341	886	1,424	951	16,082	5,883	7,815
ANB	1,021,156	430,782	73,528	4,588	8,697	226,531	36,824	77,982	69,507	93,435

Table 2 Cropping pattern

Name	Rice		Wheat		Barley		Maize		Mandua		Sava	
	Not irrigated	Irrigated	Not irrigated	Irrigated	Not irrigated	Irrigated	Not irrigated	Irrigated	Not irrigated	Irrigated	Not irrigated	Irrigated
JMT	414	74	675	58	15	–	8	–	366	–	31	–
DSL	2,033	486	2617	464	157	–	25	–	1,794	–	298	–
GHT	2,011	190	1549	172	147	–	13	–	1,775	–	280	–
KPR	302	47	1,298	39	95	–	7	–	265	–	182	–
NBR	1,471	234	1,620	224	177	–	17	–	1,299	–	338	–
GRS	1,639	194	2,993	177	208	–	62	–	1,445	–	398	–
TRL	1,440	189	730	147	58	1	20	–	1,271	–	108	–
DBL	517	23	730	25	60	–	8	–	455	–	113	–
PKR	2,514	327	3,099	283	301	–	34	–	2,218	–	579	–
OKM	1,998	415	2,230	285	290	–	53	–	1,399	–	871	–
AGM	4,329	1,243	4,159	988	159	14	71	–	3,029	–	476	–
JKL	1,876	743	3,381	1,048	532	27	34	–	1,561	–	646	–
DVP	1,249	570	4,343	355	575	13	206	–	2,968	–	3,832	11
KTN	1,553	975	4,345	820	131	6	163	–	2,435	–	3,070	5
KOT	1,389	287	1,928	285	429	–	143	–	1,182		1,280	
PRI	1,167	233	1,655	197	225	–	146	1	1,392		1,047	
KRS	928	156	1905	223	174	3	80	3	1,263		893	
ANB	26,830	6,386	39,257	5,790	3,733	64	1,090	4	26,117		144,42	16

Table 3 Cropping pattern (remaining)

Name	Urd	Masur	Pea	Arhar	Mustered		Linseed	Soya bean	Potato	
JMT	29	7	2	7	49	–	9	26	823	10
DSL	15	7	–	18	33	–	6	20	288	–
GHT	74	7	–	10	57	3	3	13	337	6
KPR	41	2	–	4	98	–	–	10	288	–
NBR	81	10	3	25	60	3	–	6	100	–
GRS	26	10	–	19	49	–	14	10	167	–
TRL	76	5	4	8	63	2	4	11	146	–
DBL	11	5	–	12	24	–	3	11	466	–
PKR	66	9	–	14	109	3	5	13	465	–
OKM		53			67		67		55	
AGM		65			73		73		22	
JKL		24			39		39		40	
DVP	146	107	16	117	101	14	61	108	114	
KTN	138	92	14	54	82	13	52	71	108	
KOT	78	31	1		4		17		12	
PRI	220	27			2		14		15	
KRS	40	37	1		2		22		14	
ANB	1,041	498	41	288	912	38	389	299	3,460	16

Table 4 Land use (ha) in the villages of Khanda Gad 1971[a]–2007[b]

Village name	Area			Forest (Van Panchayat)		
	1971	2007	Change	1971	2007	Change
Margaon	32	9.3	−22.7	–	3.6 (1952)	3.6
Shrikot Khanda	132.4	122.0	−10.4	0.8	13.6	12.8
Margadna	26.8	32.7	5.9	–	–	–
Bhitai Malla	74	73.4	−0.6		5.2 (2000)	5.2
Kaldung	92.8	80.9	−11.9	1.2	6.1 (2001)	4.9
Dhanak	25.6	26.1	0.5	–	3.0 (1985)	3.0
Gaduwa Gad	47.6	44.7	−2.9	–	–	–
Kamand	25.2	25.5	−0.3	–	5.5 (2000)	5.5
Dov	43.6	53.7	10.1	0.4	14.7 (2001)	14.3
Shiyar Malla	13.2	33.8	20.6	–	–	–
Rawat Gaon	46.4	46.6	−0.2	–	–	–
Bhitai Talla	166.4	71.9	−94.5	–	7.1 (2000)	7.1
Total	726	620.2	−105.8	2.4	58.8	56.4

Source [a] Census handbook 1971, [b] Patwari circle, Srinagar and Pauri

Table 5 Land use (ha) in the villages of Khanda Gad 1971[a]–**2007**[b]

Village name	Irrigated			Un-irrigated		
	1971	2007	Change	1971	2007	Change
Margaon	–	1.9	1.9	13.2	0.9	−12.3
Shrikot Khanda	–	5.0	5.0	33.2	36.1	2.9
Margadna	0.8	1.2	0.4	21.6	28.3	6.7
Bhitai Malla	–	3.7	3.7	17.6	46.1	28.5
Kaldung	0.8	0.7	−0.1	22.8	58.0	35.2
Dhanak	2.4	2.6	0.2	6.4	13.3	6.9
Gaduwa Gad	2.4	1.9	−0.5	15.2	32.3	17.1
Kamand	1.6	1.2	−0.4	14	14.5	0.5
Dov	–	–	–	23.2	25.2	2.0
Shiyar Malla	–	1.3	1.3	2.8	1.3	−1.5
Rawat Gaon		0.4	0.4	17.6	32.4	14.8
Bhitai Talla	4.0	3.3	−0.7	74.0	18.8	−55.2
Total	14.4	23.2	8.8	261.6	307.2	45.6

Source[a] Census handbook 1971, [b] Patwari circle, Srinagar and Pauri

Table 6 Land use land cover changes in cluster of villages in Kewer Gadhera sub-watershed between 1971[a] and 2007[b]

Village name	Area (ha)			Forest		
	1971	2007	Change	1971	2007	Change
Kewer Talla	32	31.1	−0.9	1.2	–	−1.2
Kewer Malla	47.5	45.4	−2.1	1.2	–	−1.2
Bhagoti	100	104.5	4.5	1.4	2.6	1.2
Ratni	37.6	37.8	0.2	–	1.7	1.7
Keshwan	50	25.5	24.5	–	–	–
Gadseer	156	171.6	15.6	4	7.5	3.5
Bunga	112	129.4	17.4	12	68.6	56.6
Jhijodi	204	162	−42	40	10.7	−29.3
Ali	11.5	11.1	−0.4	0.4	5.8	5.4
Leguna	14	19	5	2	7	5
Bedula	71.6	71	−0.4	4.3	22.4	18.1
Chirona	26.8	27	0.2	13	19.7	6.7
Kaub	231.6	228.6	3	20.2	49.7	29.7
Naini	22.8	22.6	−0.2	5.8	5.7	−0.1
Swan Malla	2	28.2	26.2	6	5.2	−0.8
Swan Talla	56.4	58.3	1.9	4	9.8	5.8
Kimoli	242	247.9	5.9	40.2	46.6	6.4
Total	14,17.8	1,421	58.4	155.7	263	107.5

Source [a] Census hand book 1971, [b] Patwari circle, Narainbagar

Table 7 Land use land cover changes in cluster of villages in Kewer Gadhera sub-watershed between 1971[a] and 2007[b]

Village name	Irrigated			Un-irrigated		
	1971	2007	Change	1971	2007	Change
Kewer Talla	0.8	–	−0.8	17.4	17.9	0.5
Kewer Malla	–	–	–	18.4	18.3	−0.1
Bhagoti	–	–	–	40	64.8	24.8
Ratni	–	–	–	18.8	19.2	0.4
Keshwan	–	–	–	24	7.2	16.8
Gadseer	–	–	–	64	65.3	1.3
Bunga	–	–	–	40	44.2	4.2
Jhijodi	–	–	–	46	72.3	26.3
Ali	–	0.1	–	4.4	4.3	−0.1
Leguna	–	0.8	–	8	6.1	−1.9
Bedula	–	0.7	–	24.8	24.5	−0.3
Chirona	–	–	–	5	5.1	0.1
Kaub	–	–	–	93	95.9	2.9
Naini	–	–	–	9.2	8.4	−0.8
Swan Malla	–	–	–	6	9.7	3.7
Swan Talla	–	–	–	30	31.4	1.4
Kimoli	148	–	–	74	87.9	13.9
Total	148.8	1.6	−0.8	523	582.5	93.1

Source [a] Census handbook 1971, [b] Patwari circle, Narainbagar

Table 8 NTFPs (including H and MPs) information collected from the forests in selected districts viz. Chamoli (Garhwal region) and pithoragarh (Kumaon region) (Uttarakhand State)

S.	Botanical names (families)	Local names	Habits	Parts used	Season of collection	Uses
1.	Aconitum atrox Stapf. (Ranunculaceae)	Vatsnam/Meetha Vish	Herb	Roots/Rhizome	September–October	Whooping cough and asthma. Ayurvedic and
2.	Aconitum falconeri Stapf. (Ranunculaceae)	Meetha Vish/ Vatsnam	Herb	Roots/Rhizome	September–October	Whooping cough and asthma. Ayurvedic and Yunani medicines; roots used for nervous and digestive diseases; also for rheumatism and fevers
3.	Heterophyllum Wall. ex Royle (Ranunculaceae)	Atis Ativisha	Herbs	Roots/Rhizome	September–October	For cough, strength and as tonic, astringent, stoma chic. Used in powder form aphrodisiac roots considered a valuable febrifuge and bitter tonic. Also used for hysteria, throat infection and diabetes
4.	Aconitum laciniatum Stapf. (Ranunculaceae)	Murilla	Herb	Roots	September–October	Root extract used in cure of ailment locally
5.	Acorus calamus L. (Araceae)	Bach/Vacha/Safed bach	Herb	Roots, root-stem transition	September–October	Stimulant, carminative, tonic and stoma chic, useful in dyspepsia, flatulence, loss of appetite, hysteria, epilepsy and insanity. Also used for glandular and abdominal tumors; alcoholic extract of the rhizome has sedative and analgesic properties
6.	Adiantum venustum G. Don (Adiantaceae)	Hansraj	Herb	Leaves and whole plant	October–March	Skin diseases and treatment of bronchitis. Also used against tumors, biliousness, inflammatory diseases of the chest and ophthalmic
7.	Aesculus indica Colebr. ex Camb. (Hippocastanaceae)	Pangan/Pangar/ Bankhor/ Pangla	Tree	Fruit and seeds	July–August	In stomach disorder of horses, arthritis. In allopathic medicines, cure for skin diseases; roots used in leucorrhoea; bark made into a paste and applied to dislocated joints
8.	Allium stracheyi L. (Liliaceae)	Jambu	Herb	Leaves and flowers	September–October	Used as spice for flavoring and spluttering food items
9.	Angelica glauca Edgew. (Apiaceae)	Gandrayan/Chora	Herb	Roots/Seeds	September–October	As spice and root oil has demand in international market for scenting the food items. Also used as a cardio active, carmi-native, expectorant, diaphoretic

(continued)

Table 8 (continued)

S.	Botanical names (families)	Local names	Habits	Parts used	Season of collection	Uses
10.	Artemesia maritima L. (Asteraceae)	Pati/Chamur/ Kirmala/ Afsanthin	Shrub	Leaves flowers	September–October	Leaves scented. In preparation of Ayurvedic medicine Naglona. Possesses antihelmi-nitic, antiseptic properties. Also used for asthma and psychological diseases and is recommend-ed for chronic fever and swellings; yields an essential oil called wormwood oil which has a tonic effect on digestive organs; also used externally for rheumatism
11.	Asparagus racemosus Willd. (Liliaceae)	Satawar/Shatawar/ Satmata/Phusar	Herb	Roots and leaves	Throughout the year	Gives cooling effect. Roots demulcent, diuretic, aphrodisiac, antispasmodic and alterative tonic. Leaves used for treatment of boils, smallpox. Fresh root juice is mixed with honey and given for dyspepsia; these also form a constituent of medicinal oils used for nervous and rheumatic complaints; roots also used as demulcent, aphrodisiac, diuretic, anti-dysenteric, and in veterinary medicines
12.	Berberis asiatica DC. (Berberidaceae)	Kilmoda, Rasaut, Daru-haldi	Bush	Roots and bark	October–March	Recommended in excess urethral discharge, eye diseases, jaundice, and fevers. Root bark used for skin diseases, eye ailments and malaria. Roots useful in healing of wounds
13.	Berberis aristata DC. (Beberidaceae)	Dar-haldi	Shrub	All parts especially root bark/stem	July–August	Malaria, jaundice, eye disease, skin and stomach diseases. In Ayurvedic and Yunani system it is used for preparation of Berberis hydrochloride
14	Berginia ciliata Sternb (Saxifragaceae)	Pashanabheda/ Shailagarbhaja/ Silphori	Herb	Rhizome	September–March	The drug Pashanabheda is extracted from dried rhizomes. It possesses astringent, tonic, antiscor-butic and laxative prop-erties. Useful in pulmo-nary infection, dysentery, ulcers, spleen enlarge-ment, cough and fever

(continued)

Table 8 (continued)

S.	Botanical names (families)	Local names	Habits	Parts used	Season of collection	Uses
15.	Cannabis sativa L. (Cannabinaceae)	Bhang/Ganja	Shrub	Seeds, dried leaves	September–October	As intoxicant, stomachic, analgesic, narcotic, sedative, pain killer, and anodyne
16.	Centella asiatica (L.) urban (Hydrocotylaceae)	Mandukparni	Herb	Leaves/Stem	September–October	Treatment of cold and useful for improving brainpower also. Used in Ayruvedic and Yunani systems of medicines as diuretic, alterative and tonic and has demand in International market also. A decoction of the whole plant is used in treatment of leprosy
17.	Chenopoduim album L. (Chenopodiaceae)	Bathua	Herb	Leaves	June–September	Appetizer and laxative
18.	Cinnamomum tamala (Buchanan-Hamilton) T. nees and eberm. (Lauraceae)	Tejpat	Tree	Leaves	October–onward	Leaves used as Tejpat or Tejpatta leaf. Useful in scabies, disease of the anus and rectum such as piles (in Ayruveda), colic and diarrhea
19.	Cinnamomum zeylanicum Blume (Lauraceae)	Dalchini	Tree	Inner bark	September–November	Bark used as Dalchini (a spice). It has a sharp hot taste, tonic; used as alexiteric, stimulant, car- minative, expectorant, aphrodisiac and for headache, etc. Used in Ayurvedic and Yunani systems of medicines
20.	Citrullus colcocynthis (L.) Kuntze (Cucurbitaceae) Ayurvedic and Yunani medicines for blood pressure, stomachache, jaundice, urinary diseases and rheumatism. Fruit and roots antidote to snake poison, extract from fruit pulp is highly effective against bacteria	Indrayan	A	perennial herb	Fruits, roots	June–July
21.	Coleus forskohlii (Willd) Briq. (Lamiaceae)	Coleus	Herb	Roots	September–October	Blood pressure and glaucoma, in allopathic medicines

(continued)

Table 8 (continued)

S.	Botanical names (families)	Local names	Habits	Parts used	Season of collection	Uses
22.	Corydaylis govaniana Wall. (Fumariaceae)	Bhutkeshi/ Jatamansh	Herb	Root	September–October	Insecticide, strength, blood purifier. Roots locally used as incense; also recommended against syphilis and coetaneous affections
23.	Datura stramonium L. (Solanaceae)	Dhatura	Herb	Flower, leaves, fruit, seeds	October–November	Sleepiness, intoxicant, skin diseases, fever leprosy and asthma. In Ayurvedic, allopathic and Yunani medicines; alkaloid atropine used as stimulant for the central nervous system and as sulphate for dilating pupil
24.	Delphinium denudatum Wall. (Ranunculaceae)	Nirbishi	Herb	Root	September–November	The root is bitter and considered as stimulant, alternative and tonic. Also useful as tonic for toothache and an adulterant for aconite
25.	Didymocorpus pedicellata R. Br. (Gesneriaceae)	Patthar Phori	Herb	Leaves and bud	–	As scent and in Ayurvedic medicines as remedy for kidney and bladder
26.	Dioscorea bulbifera L. (Dioscorcaceae)	Vangenthi/Gainthi/ Ratalu	Herb	(Climb-er)	Tuber	October – January
	For strength and nutrition in Ayrvedic medicine e.g. Chyvanparash. Tuber used in piles, syphilis and applie to ulcers					
27.	Dioscorea deltoidea Wall. ex Kunth. (Dioscoreaceae)	Kathparun/Kithi	Climb-er	Tuber (stem)	October–January	In cortisone hormones and Allopathic medicines, tubers yield steroidal sapogenins, which are sources of oral contraceptives
28.	Diploknema butyracea (Roxb.) H.J. Lam. (Sapotaceae)	Chyura (Indian Butter Tree)	Tree	Fruit	June–July	Edible oil used as butter and in preparation of chocolate and hair pomade
29.	Emblica officinalis Gaertn. (Euphorbiaceae)	Amla/Aonla	Tree	Flower, Fruits, roots	October–November	Fruits used as diuretic, laxative and for treatment of jaundice, cough stomach disorders, anemia and eye ailments; rich source of vitamin c; fruits have anti-biotic activity against a variety of micro-organisms

(continued)

Table 8 (continued)

S.	Botanical names (families)	Local names	Habits	Parts used	Season of collection	Uses
30.	Ephedra gerardiana Wall. (Ephedraceae)	Soam	Shrub	Twigs	August	Cure of syphilis (genital diseases), asthma, cold, hay fever and rashes of allergic origin; nasal spray is used against inflammation of mucous membrane
31.	Fritillaria roylei Hook. (Liliaceae)	Kakoli	Herb	Root tuber	October–November	In Ayrvedic and Yunani medicines roots used for healing of wounds, corms are used for asthma and bronchitis.
32.	Hebenaria comme-linifolia wall. ex Lindl. (Orchidaceae)	Vridhi	Herb	Roots	June–October	Used in Ayrvedic preparation as chyavanprash and for strength as a source of salep
33.	Hedychium spicatum Buchanan-Hamiltonex Smith (Zingiberaceae)	Kapur kachari/ Kapoor kesri/ Ban Haldi	Herb	Rhizomes	October–December	Stomachic, carminative, tonic, stimulant, Useful in liver complaints, vomiting, diarrhea, inflammations and pain as well as snakebite
34.	Indigofera pulchella Roxb. (Fabaceae)	Sakina/Hakna	Shrub	Roots	May–June	Used for treatment of cough and chest pain
35.	Juglans regia L. (Juglandaceae)	Akhrot	Tree	Fruit, bark and leaves; roots	November–January	Bark possesses insecti-cidal properties, toothpaste making, fruit edible; leaves astringent, tonic and anathematic
36.	Juniperus communis L. (Cupressaceae)	Haubera/Aaraar	Shrub or small tree	Leaves and fruit	September–October	Yields essences, scent and havan material. Fruit used as Hauber in Ayruvedic, also used for flavouring gin
37.	Adhatoela vaxiea Nees (Acanthaceae)	Adulasa/Arusa	Peren-nial herb	Leaves	June–September	In cough as expectorant bronchitis and other respiratory ailments. In allopathic, Ayurvedic and Yunani medicines; leaves also possess anthelmintic properties
38.	Lyonia ovalifolia Wall. Drude (Ericaceae)	Ayar	Tree	Leaves and buds	January–February	Insecticidal used in skin ailments. Also in folklore medicine in Kumaon and Garwal, traditional treatment of diseases

(continued)

Table 8 (continued)

S.	Botanical names (families)	Local names	Habits	Parts used	Season of collection	Uses
39.	Macrotomia benthami A. DC. (Boraginaceae)	Ratanyot or Laljadhi Kashimiri Goozaban	Herb	Root and root bark	September–October	Roots paste mixed with milk used for healing of internal wounds and soars. Good for vomiting and digestion. As colour-ing agent in spices; aqueous extract of flowering shoots is used in making sherbet and jams that are recommended for troubles of tongue and throat and for cardiac disorders
40.	Mallotus philippensis (Lamk.) Muell. – Arg. (Euphorbiaceae)	Rohini/Kamala/roli/Sindhuri	Tree	Fruit (oil glands and hair on fruits) seeds	March–May	Bitter, possesses wormicidial and insecticidal properties, fruits yield a red dye used as an anathematic and for coetaneous affections; seed oil is used for making hair fixers and ointments
41.	Mentha longifolia (L.) Huds. (Lamiaceae)	Jungli Pudina	Perre-nial herb	Leaves/ flowers	September–October	Carminative, stimulant and cooling medicines
42.	Myrica esculenla Buch.-Ham. ex D. Don (Myricaceae)	Kafal, Katfal	Tree	Bark	September–October	Bark carminative Ayur-vedic and Yunani system of medicine for cough, dysentery and dieresis; chewed for toothache
43.	Nardostachys jatamansi (D. Don) Dc. (Valerianaceae)	Jatamansi/Balchhar	Herb	Rhizome	September–October	Stimulant, strength as antiseptic, epilepsy and intestinal pain, source of an essential oil supposed to improve hair growth and black colour; its tincture is use for epilepsy and hysteria; also used as a laxative for improving urination, menstruation and digestion

(continued)

Table 8 (continued)

S.	Botanical names (families)	Local names	Habits	Parts used	Season of collection	Uses
44.	Orchis latifolia L. (Orchidaceae)	Hathandi/ Salammishri/ Salap	Herb	Stem, root	September–October	Tonic for digestive system, Cough, heart ailment and healing agent of wound in Ayurvedic and Yunani System of medicine
45.	Paris polyphylla Sm. (Liliaceae)	Satua	Herb	Root and tuber	October–March	Used in Ayurvedic medicine. The rhizome is possessing antithelmintic properties and also used as tonic
46	Picrorhiza kurrooa Royal ex Benth. (Scrophulariaceae)	Kutki, Katki	Herb	Root	September–October and May–June	Roots are used for treatment of fever, dysentery scorpion bite, eczema, catharic, stoma chic, purgative
47.	Pinus roxburghii Sarg. (Pinaceae)	Chir	Tree	Stem	Throughout the year	Resin extract is used in manufacture of turpentine oil, varnish and paints industry. Also mixed in artificial essences, camphor and scent preparations
48.	Podophyllum hexandrum Royle (Podophyllaceae)	Bankakri/Papda	Herb	Roots	September–October	Liver ailment, digestive and purgative; the drug also checks growth of cancerous cells
49.	Polygonatum multiflorum (L.) All. (Liliaceae)	Mahameda	Herb	Rhizomes	September–October	Ayrvedic medicine especially in Chavanprash; also used for treatment of piles and discolouration of skin resulting from blows
50.	P. verticillatum All. (Liliaceae)	Meda/Mitha dudhia	Herb	Roots	September–October	Ayrvedic medicine especially in Chavanprash
51.	Potentilla fulgens Hook. (Rosaceae)	Vajradanti	Peren-nial herb	Root	September–March	Liver related diseases; roots used in diarrhoea
52.	Prunus cerasoides D. Don (Rosaceae)	Padam	Tree	Bark and fruit	November–January	Bark in Ayurvedic medicine and fruit for colouring the medicines. The stem is bitter, acrid, antipyretic, refrigerant, causes flatulence, cures leprosy, hallucinations, burning of the body, etc.
53.	Rheum emodi Wall. ex Meissn. (Polygonaceae)	Dolu/uvan tantora	Herb	Root, root-stem transition zone	September–November	For strength and also ingredient in digestive medicines; powdered roots also used in healing ulcers
54.	Rhododendron arboreum Sm. (Ericaceae)	Burans	Tree	Flower	March–June	Flowers yields juice used in making squash for cold drinks; also used in diarrhea and dysentery
55.	Rhus succedanea L. (Anacardiaceae)	Kakrasingi	Shrub	Insect galls	September–October	As tonic, expectorant, cough, phthisis, asthma, fever, want of appetite and irritability of stomach; galls are astringent and expectorant

(continued)

Table 8 (continued)

S.	Botanical names (families)	Local names	Habits	Parts used	Season of collection	Uses
56.	Rubia cordifolia L. (Rubiaceae)	Manjishtha, Manjit	Herb (climb-er)	Roots/fruit		Used in Ayurvedic medicines for rheumatism; roots yield a red dye. Fruit, useful in hepatic obstruction
57.	Schima laureola(DC.) Choisy(Theaceae)	Nettpati/Kasturpatti	Herb	Leaves	September–October	Incense and as scent
58.	Stephania glabra (Roxb.) Miers (Menispermaceae)	Gindaru	Climbi-ng shrub	Roots/tubers	September–January	Burn and blood pressure; tubers are used in pulmonary tuberculosis, asthma and dysentery
59.	Swertia chirayita (Roxb. ex. Flem.) Karst. (Gentianaceae)	Chirayita	Herb	Whole plant	September–October	Ingredient in medicines of chronic fever and stomach ailments; tonic in bronchial asthma, anemia, liver disorders and also laxative
60.	Symplocos paniculata Miq. (Symplocaceae)	Lodh	A large shurb or tree	Bark	February–March	Eye ailment, coolant gum ailments. Useful in bowel complaints, diarrhea and ulcers, etc.
61.	Syzygium cuminii (L.) Skeels (Myrtaceae)	Jamun, Jambu	Tree	Bark, fruit, seeds	April–June.	Bark bitter, used in bleeding asthma and healing of wounds, fruits edible; seeds recommended for diabetic patients
62.	Taxus baccata L. ssp. wallichiana (Zucc.) Pilger (Taxaceae)	Thuner, Birumi, Dhurai, Talishpatra	Tree	Leaves and bark	May–November	Asthma, epilepsy, insanity. Bark is used for Ayurvedic and Yunani medicines against headache, giddiness, diarrhea and biliousness
63.	Thalictrum foliolosum DC. (Ranunculaceae)	Pinjari, Mamira	A tall herb	Roots	September–October	Used as a purgative, diuretic, febrifuge and dyspepsia in Ayurvedic medicines and ophthalmic as well as for making kajal sticks
64.	Thymus serpyllum L. (Lamiaceae)	Banajwain	Herb	Whole plant seeds especially	July–August	Poor eyesight, liver and stomach ailments, urinary problem, Seeds have insecticidal properties. Oil used in toothache and as a spice
65.	Valeriana jatamansi Jones (Valerianaceae)	Samoy/Samau/ Mushkbala	Herb	Roots and fruits	September–October	As stimulant and antiseptic; also used in making ince-nse sticks and scents; for hair oils and perfumes and medicines for hysteria and nervous problems

(continued)

Table 8 (continued)

S.	Botanical names (families)	Local names	Habits	Parts used	Season of collection	Uses
66.	Viburnum cotinofolium D. Don (Caprifoliaceae)	Karra/Ghinwa	Shrub	Bark	October–November	Ayurvedic medicines used as Viburnum bark for menorrhagia and metrorrhagia
67.	Viola odorata L. (Violaceae)	Banfsha/Vanafsa	Herb	Roots flower and fruit	March–May	Roots antihelmintic, antipyretic, febrifuge. Flowers emollient, demulcent, used in biliousness and lung troubles; flowers used in Ayurvedic and Unani systems for several skin, eye and ear diseases; also used as blood purifier
68.	Woodfordia fruticosa (L.) Kurz.	Santha/Dawi	Bush	Tuber	January–March	In Ayurvedic and Yunani medicines, solutions / liquids, bark extract is used as a colouring agent and for tanning
69.	Zanthoxylum armatum DC. (Rutaceae)	Timur/Tejphal/ Darmar	Tree	Stem bark and seeds	November–January	Teeth and gum ailment including pyorrhea; Treatment of scabies seeds used as a tonic; bark, fruits and seeds are extremely used in medicines as carminative, stoma chic and anathematic; essential oil from seeds used for making tooth powders
70.	Tanacetum nubigenum Wall. ex DC. (Asteraceae)	Dhup/Gugal	Herb	Roots	September–October	In pain and fever, as an insecticide; gum resin is used as incense

Source Natural Resources India Foundation: Pilot Study on Mechanism for sustainable development and promotion of herbal and medicinal plants in the state of Uttarakhand, SER division planning commission, Govt. of India

Table 9 Inventory of medicinal by HABITS occurring in Uttarakhand

Botanical names	Vernacular names	Families	Habit	Part used	Status
Acacia concinna DC.	Shikakai	Mimosaceae	Tree	Pods/leaves	Endangered
Cassia fistula Linn.	Amaltas	Caesalpianaceae	Tree	Fruit pulp	Cultivated
Cedrus deodara (Roxb. ex Lamb.) G. Don	Deodar	Pinaceae	Tree	Wood oil	Common
Hydnocarpus laurifolia (Dennst.) Sleumer	Rowti/Taurella	Flacourtiaceae	Tree	Seeds	Rare
Mesua ferrea L.	Nagkesar	Clusiaceae (Guttiferae)	Tree	Flowers	Cultivated
Pistacia integerrima Stewart ex Brandis	Kakarasinghi	Pistaciaceae	Tree	Leaf galls/wood	Scarce
Prunus cerasoides D. Don	Padam	Rosaceae	Tree	Heartwood/Fruit / endocrp	Common
Rhododendron arbore -um Sm.	Burans	Ericaceae	Tree	Flowers/Bark/tender leaves	Common
Strychnos nux-vomica L.	Karaskara/Kuchla	Strychnaceae	Tree	Seeds	Scarce
Symplocos racemosa Roxb.	Lodhra	Symplocaceae	Tree	Bark	Scarce
Taxus baccata Linn. sp wallichiana (Zucc.) Pilger	Thuner/Talishpatra	Taxaceae	Tree	Leaves/Bark	Scarce
Zanthoxylum armatum DC.	Tejphal/Darmar	Rutaceae	Tree	Wood/seeds	Scarce
Aconitum atrox Stapf	Mithavish	Ranunculaceae	Herb	Roots	Threatened
A. heterophyllum Wall. Ex Royle	Ativish	Ranunculaceae	Herb	Roots	Threatened
A. violaceum Jacq.	Patish/Dudhia vish	Ranunculaceae	Herb	Roots	Endangered
Acorus calamus L.	Vacha/vach	Araceae	Herb	Rhizomes	Common
Anacyclus pyrethrum DC.	Akarkara	Apiaceae	Herb	Roots	Scarce
Angelica archangelica L.	Chora	Apiaceae	Herb	Roots/seeds	Scarce
A. glauca Edgew.	Gandrayani	Apiaceae	Herb	Roots/fruits	Endangered
Apium graveolens	Ajmada/Solari	Apiaceae	Herb	Roots	Cultivated
Artemisia maritima L.	Kirmala	Asteracea	Herb	Whole plant	Common
Asparagus adscendens Roxb.	Shatavar	Liliaceae	Herb	Roots	Cultivated
Bacopa monnieri (Linn.) Pennell	Brahmi	Scrophulariaceae	Herb	Whole plant	Common

(continued)

Table 9 (continued)

Botanical names	Vernacular names	Families	Habit	Part used	Status
Berginia ciliata Sternb	Pashanbhed/Silphori	Saxifragaceae	Herb	Roots	Rare
Cannabis sativa Linn.	Bhang	Cannabinaceae	Herb	Leaves	Abundant
Centella asiatica (L.) Urban	Mandukparni	Hydrocotylaceae	Herb	Whole plant	Common
Citrullus colocynthis (L.) Kuntze	Indrayan	Cucurbitaceae	Herb	Whole plant	Cultivated
Colchicum luteum Baker	Hirantutiya	Liliaceae	Herb	Whole plant	Cultivated
Coptis teeta Wall.	Mamira/Mishmi	Ranunculaeae	Herb	Rhizomes	Rare
Curculigo orchiodes Gaertn.	Kalimusli	Hypoxidaceae	Herb	Roots	Scarce
Curcuma amada Roxb.	Amahaldi	Zingiberaceae	Herb	Roots/rhizomes	Common
Cyperus rotundus Linn.	Motha	Cyperaceae	Herb	Rhizomes	Common
Dactylorhiza hatagirea (Don.) Soo	Hathjari	Orchidaceae	Herb	Roots	Rare, now cultivated
Delphinium denudatum Wall.	Nirbishi	Ranunculaceae	Herb	Roots	Rare
Didymorarpus pedicellata R.	Pathar laung/	Gentianaceae	Herb	Leaves	Rare
Br. Dioscorea deltoidea Wall. ex Kunth	Patthar Phori Gainthie	Dioscoreaceae	Herb	Tubers	Cultivated
Eulophia campestris Wall.	Salibmisri	Orchidaceae	Herb	Roots/tubers	Endangered
Fumaria officinalis L.	Pitt Papara	Fumariaceae	Herb	Whole plant	Common
Gentiana kurroo Royle	Nilkanthi, Kurn/Kutki	Gentianaceae	Herb	Whole plant	Common
Gloriosa superba Royle	Kalihari	Liliaceae	Herb	Tubers/leaves	Rare
Hedychium spicatum Buch.– Ham. Smith	Kapur kachri/Sitruti	Zingiberaceae	Herb	Rhizomes	Scarce
Hyoscyamus niger L.	Khurasani ajvayan	Solanaceae	Herb	Leaves	Cultivated
Mucuna pruriens (L.) DC.	Konch/Kawnch	Fabaceae	Herb	Seeds/fruits	Common
Nardostachys jatamansi (D. Don) DC	Jatamasi/Balchhar	Valerianaceae	Herb	Rhizomes	Endangered
Nelumbo nucifera Gaertn.	Kamal	Nelumbonaceae	Herb	Stem/Seeds	Common
Orchis latifolia L.	Salammisri/Salap	Orchidaceae	Herb	Leaves	Rare
Papaver somniferum L.	Aphim	Papaveraceae	Herb	Capsules	Cultivated
Plantago ovata Forsk	Isabgol	Plantaginaceae	Herb	Seed husk	Common

(continued)

Table 9 (continued)

Botanical names	Vernacular names	Families	Habit	Part used	Status
Podophyllum hexandrum Royle	Bankakri/Vankakri	Podophyllaceae	Herb	Roots	Endangered
Polygala chnensis Ham.	Miragu	Polygalaceae	Herb	Roots	Common
Rheum emodi Wall. ex Miers.	Archu/Revandchini	Polygonaceae	Herb	Roots	Endangered
Rubia cordifolia Linn.	Manjeeth/Manjeet	Rubiaceae	Herb	Roots	Common
Saussurea costus (Falc.) Lipsch	Kuth	Asteraceae	Herb	Roots	Common/ Cultivated
S. obvallata Wall. ex C.B. Clarke	Brahmakamal	Asteraceae	Herb	Roots	Common
S. lappa C.B. Clarke	Kuth	Asteraceae	Herb	Roots	Common
Swertia chirayita (Roxb. ex Flem.) Karst	Chirayita	Gentianaceae	Herb	Whole plant	Roadside plantation
Tephrosia purpurea (L) Pers.	Sharepunkha	Fabaceae	Herb	Leaves	Common/ Cultivated
Tinospora cordifolia Miers ex Hook F. and Thoms.	Giloi/Gurach	Menispermaceae	Climbing - shrub	Whole plant	Common
Urginea indica (Roxb.) Kunth.	Jangli piyaz/Banpiaj	Liliaceae	Herb	Whole plant	Common
Valeriana hardwickii Wall.	Sugandhbala/Tagger	Valerianaceae	Herb	Roots	Scarce
V. jatamansi Jones	Tagar/Sameva	Valerianaceae	Herb	Rhizomes	Rare
Viola odorata L.	Vanafsa/Banafola/ Banafsa	Violaceae	Herb	Flower/fruit	Common/ Cultivated
Berberis aristata DC.	Daruharidra	Berberidaceae	Shrub	Roots/bark	Rare
B. asiatica Roxb.	Kilmora/kingora	Berberidaceae	Shrub	Roots/bark	Rare
Calotropis giganteum Linn.	Madder/Madar	Asclepiadaceae	Shrub	Whole plant	Common
Cephaelis ipecacuanha (Brot.) A. Rich.	Ipecac	Rubiaceae	Shrub	Whole plant	Rare
Ephedra gangetica Wall.	Som	Ephedraceae	Shrub	Stem	Scarce
E. gerardiana Wall.	Som/Asmania	Ephedraceae	Shrub	Whole plant	Rare
Glycyrrhiza glabra Linn.	Mulethi	Fabaceae	Shrub	Roots/stems	Cultivated

(continued)

Table 9 (continued)

Botanical names	Vernacular names	Families	Habit	Part used	Status
Juniperus communis Linn.	Dhup lakri	Cupressaceae	Shrub	Whole plant	Common
Plumbago indica L.	Lal Chitra	Plumbaginaceae	Shrub	Roots	Grown in gardens
P. zeylanica L.	Chitter/Chitrak	Plumbaginaceae	Shrub	Roots	Rare, now Cultivated
Pueraria tuberosa (Roxb. ex Willd.) DC.	Vadarikand, Siari	Fabaceae	Shrub	Tuber/ leaves	Scarce
Tylophora indica (Burm. F.) Merr.	Anantmool	Asclepiadaceae	Climbers	Leaves/roots	Endangered
Uncaria gambier (Hunt.) Roxb.	Math-kuntha, Kath-Kuntha	Rubiaceae	Climbing, Shrub	Leaves/branches	Rare
Withania somnifera (L.) Dunal	Ashagandgha/Asgandha	Solanaceae	Shrub	Roots	Rare
Dioscorea bulbifera L.	Ratalu/Gainthie	Dioscoreaceae	Climbing	Aerial tubers	Common
Entada phaseoloides (L.) Merr.	Gadbigh/Gadbich Birabi/ Chian	Mimosaceae	Huge climbing shrub	Seeds/stem/bark	Common
Smilax glabra Roxb.	Bari Chobcheeni	Smilacaceae	Climber	Roots/bulb	Common
Trichosanthes dioica Roxb.	Potala/Parwal	Cucurbitaceae	Climber	Fruits	Cultivated

Source Natural Resources India Foundation: Pilot Study on Mechanism for sustainable development and promotion of herbal and medicinal plants in the state of Uttarakhand, SER division planning commission, Goverment of India

Table 10 Main socio-economic factors behind responsible for traditional crops/landraces/germplasm (After Bisht et al. 2006)

S. No.	Important factors responsible for agro-biodiversity loss	Process of change and implication
1.	Change in cropping patternsdue to economicconsiderations	The farmers in the region are involved in diverse livelihood options as cultivation of crops, livestock, forestry, etc. Many of the traditional crops are grown under marginal conditions and often provide low yield and extremely low income, forcing the farmers to undertake other activities, for example, replacement of mixed cropping to monocropping, cultivation of improved strains bringing about more uniformity in crop species and switching over to cash crops. Mono-cropping and uniformity results in increased vulnerability to pest epidemics and consequent loss of biodiversity. Besides, a significant proportion of the traditional agricultural land has been brought under cash crops or o.—season vegetables. This has adverse implications on traditional agro-ecosystemsand traditional agro-biodiversity of the region has shrunk over the time
2.	Population growth and landfragmentation	The human population has increased over time. The land fragmentation and insufficient crop yield due to high land: man ratio and low output: input ratio of traditional crops compelled farmers to consider other options for livelihood
3.	Lack of traditional knowledge	Since there is no systematic documentation of ethno-medicinal uses of traditional landraces and the traditional underutilized crops, the younger generation is unaware about the distinctive properties of the landrace diversity. Lack of this knowledge often leads to discontinuation of cultivation of some of these landraces which are of high nutritional value to them. This kind of knowledge is, however, very much essential for value addition to local landrace diversity and also in IPR protection
4.	Out-migration	Migration of people to plain areas for o.—farm jobs and reduced interest in traditional agriculture
5.	Change in food habits	Yield potential of most of the traditional crops has been stable for the last 2–3 decades. The food shortage problem is because of population growth, change in food habits (increasing preference for wheat and rice as staples), reduction in crop diversity and net sown area
6.	Social values	Local socio-cultural integration has decreased. Social institutions such as community participation in natural resource management for agriculture, and seed and labour exchange systems are disappearing fast leading to weakening of agricultural management

Table 11 Changes in faming system

Crops/cropping season Summer season crop	Replacement crop	% decline in traditional crops
Panicum miliaceum (Cheena)	High Yielding rice varieties	65.5
Oryza sativa (Traditional land races)	High Yielding rice varieties	100
Avena sativa (Jai)	Potato	78.5
Fagopyrum tataricum (Phaphar)	Potato and rajma	82.5
F. esculentum (Oggal)	Rajma	92.7
Parilla frutescense (Bhangjeera)	Soyabean	100
Setaria italica (Kauni)	Soyabean	65.2
Eleusince coracana (Koda)	Soyabean and amaranth	36.5
Echinochloa frumentacea (Jhangora)	Pigeon Pea	72
Macrotyloma uniflorum (Gahat)	Soyabean and amaranth	70
Vigna sp. (Mass)	Pigeon pea and amaranth	100
Winter season crops		
Triticum aesativum (traditional landraces of wheat) + Brassica (sarson)	High yielding wheat varieties	100
Hordeum himalayens (Owa)	Potato, amaranth and rajma	95
Hordeum vulgare (Jau)	Improved mustard varieties	84.3

Source Nautiyal, S., Bisht, V., Rao, K. S., & Maikhuri, R. K. (2008). The role of cultural values in agrobiodiversity conservation: A case study from Uttarakhand, Himalaya. *Journal of Human Ecology, 23*, 1–6

Table 12 Traditional commercial crops

English name	Vernacular name	English name	Vernacular name
Amaranth	Chaulai	Soybean	Bhatt
Amaranth	Chuwa/Marcha/Ramdana	Soybean	Kala Bhatt
Pigeon pea	Tor	Soybean	Soybean
Pig-weed	Bethu	Naked barley	O-wa-jau
Taro	Pindalu/Kuchain	Horesgram	Gahat
Buckwheat	Oggal	Potato	Alu
Buckwheat	Phaphar	Mat bean	Bhirnga
Maize	Mungri	Adjuki bean	Rains
Rice bean	Bhotia	Black gram	Urd
Zinger	Adrak	Cow pea	Sonta
Poppy	Post	Hog-millet	Cheena Jakhiya

Table 13 Threatened medicinal plants of Uttarakhand

Name of the plant	Family	Local/trade name	Status	General distribution	Distribution in Uttarakhand
Part I					
Nardostachys grandiflora DC.	Valerianceae	Jatamansi	Vulnerable	India (Garhwal to Sikkim) Bhutan, Nepal and W. Tibet	Garhwal and Kumaon 3300–4500 m
Fritillaria toylei Hook.	Liliaceae	Kashir kokoli	Critically rare	India (Kashmir to Kumaon) China (W. Tibet)	Garhwal and Kumaon 2600–4000 m
Onosma bracteatum Wall.	Boraginaceae	Gojhiva	Rare	India (Garhwal to Kumaon) Nepal	Garhwal and Kumaon 3300–4500 m
Podophyllum hexandrum Royle.	Podophyllaceae	Ban kakri	Endangered	India (Kashmir to E. Himalaya) Nepal	Garhwal and Kumaon 3000–4000 m
Rheum australe D Don = (*Rheum emodi* Wall).	Polygonaceae	Revan Chini	Vulnerable	India (H.P. to Sikkim) Nepal, Bhutan, S.W.	Garhwal and Kumaon 3300–4000 m
Picrorhiza kurrooa Royle	Crophulariaceae	Kutki	Endangered	India (Garhwal and Kumaon) Nepal, Bhutan, China and Mayanmar	Garhwal and Kumaon 3300–4000 m
Swerita chirayita (Roxb. Ex Flem). Karsten	Gentianceae	Chirayita	Critically Rare	India (Kashmir to Sikkim and Assam), Nepal and Bhutan	Garhwal and Kumaon 1800–2500 m
Dactylorhiza hatagirea (D. Don) Soo = (*Orchis latifloia* L.)	Orchidaceae	Salam panja	Endangered	India (Kashmir to Sikkim), Nepal, Bhutan, China, Pakistan	Alpine region of Garhwal and Kumaon 3000–4000 m
Angelica glauca Edgew	Apiaceae	Choraka, Choru	Vulnerable	India (Kashmir to Garhwal and Kumaon	Garhwal 2700–3600 m
Selinum elatum (Edges) Hiroe = (*Selinum tenuifolium* Wall. Ex Cl.)	Apiaceae	Murva		India (Kashmir to Sikkim and Assam) Nepal and China	Garhwal and Kumaon 3000–4300 m
Gentiana kurroo Royle	Gentianaceae	Karua/True Kutki	Rare,	Endangered	India (Kashmir to Kumaon)
Garhwal and Kumaon					
Part II					
Aconitum heterophyllum wall.	Ranunculaceae	Ativisha/ Ateesh	Vulnerable	India (Kashmir to Sikkin), Nepal	Garhwal and Kumaon 3600–4500 m

(continued)

Table 13 (continued)

Name of the plant	Family	Local/trade name	Status	General distribution	Distribution in Uttarakhand
Lilium polyphyllum D. Don	Litiaceae	Kakoli	Rare	India (Kashmir to Kumaon), Afganistan and Nepal	Garhwal and Kumaon 2200–3500 m
Malaxis muscifera (Lindl.) Kuntze = *Microstylis muscifera* (Lindl.) Rdl.	Orchidaceae	Jivak	Rare	India (Kashmir to Sikkim), Nepal, Bhutan, Myanmar, and W. China	Garhwal and Kumaon 2600–4000 m
Celastrus paniculatus Wild.	Celastraceae	Jyotismati	Vulnerable	India (H. P. to Assam UP to 750 m from plains, Nepal)	Outer Himalaya tropical forest of Uttaranchal
Curculigo orchioides Gaertn.	Hypoxidaceae	Talmulika	Vulnerable	India (Jammu to Assam and South India), Nepal, Sri Lanka, Japan, Malaysia and Australia	Tropical forests up to 500 m
Taxus wallichiana Zucc. (=*taxus baccata* L.)	Taxaceae	Talish/ Thuner	Critically Rare	India (Western to Eastern Himalaya), Nepal, Bhutan, Myanmar, Malaysia and Indo China	Garhwal and Kumaon 2600–3800 m
Abies pindrow Royle	Pinaceae			India (Kashmir to Kumaon), East Afganistan and Nepal	Garhwal and Kumaon 2300–2700 m
Abies spectabilis (D. Don) Mirb = (*Abies webbiana* Lindl.)	Pinaceae	Aitalish		India (Kashmir to Kumaon) E. Afghanistan and Nepal	Garhwal and Kumaon 3000–4000 m

Table 14 State Government initiatives for enhancing livelihood

Objective	Scheme	Direct outlay/time period/no. of beneficiaries	Indirect outlay/period	Indirect outlay/period (2)/period
To alleviate poverty	Livelihood assistance/training and skill up gradation/revolving fund to SHGs	Self employment (annual)	Rs. 2,000/- pm after 3 years	Reduction of poverty in %age –8.25 %
		(1) Nos. of Swarozgaris of SHGs under SGSY—1,500	Rs. 1,200/- pm after 6 months	
		(2) Namada woolen activity Training—200	Rs. 1,000/- pm after 6 months	
		(3) Off season vegetable cultivation by polyhouses—556	Rs. 2,500/- pm after 6 months	
		(4) Fisheries—10	Rs. 1,000/- pm after 6 months	
		(5) Training to unemployed youth in ITI—170	Rs. 5,000/- pm after 2 Years	
		(6) Employment oriented books to unemployed youth—1,000		
To increase agricultural productivity	(1) Minor irrigation (2) Major irrigation (3) Agriculture (4) Check dams/ land Mgmt.	Irrigated area—75.38 Ha Irrigated area—75.38 Ha Improve seed distribution Checking landslides	Increase in production due to improved seed supply and irrigation	Per hectare income of Rs. 2850 in Kharifcrops and Rs. 3675 in Rabi crops
School Enrolment and in literacy	(1) Provision of books (2) Prov. of Lab Eqpts. (3) Provision of sports goods	In 244 schools In 210 schools In 60 schools	Increase in enrolment	Increase in attendance

(continued)

Table 14 (continued)

Objective	Scheme	Direct outlay/time period/no. of beneficiaries	Indirect outlay/period	Indirect outlay/period (2)/period
ent	Road and bridges construction	1st Year—2.00 km/1 bridge	To connect rural areas with health centres, block HQs and inter colleges health care dev. and population control	Increase in marketing and income
	(1) Strengthening of primary health Centres	2nd Year—15.00 km/3 bridges		
	(2) Development of health facilities	3rd Year—16.55 km/2 bridges		
	(3) Strengthening of anganwari centres	1st year—9		
		2nd year—9		
		3rd year—9		
		1st year—3		
		2nd year—2		
		1st year—3		
		2nd year—3		
		3rd year—3		
Transformers and HT/LT lines for improvement in Low Voltage supply	Sub Station 79 × 25 sq m	To address problems of increased power demand and low voltage supply	Establishment of power based small scale industry and generation of income by augmentation of employment	
	HT lines—29.8 km			
	LT lines—34.2 km			

(continued)

Table 14 (continued)

Objective	Scheme	Direct outlay/time period/no. of beneficiaries	Indirect outlay/period	Indirect outlay/period (2)/ period
Tourism development	Development of rural tourism, trekking and ancient routes	Developing tourism to generate alternate means of livelihood and thus augment earnings	To attract both national and international tourists for generating employment and increased income	
Jatropha plantation	Van Panchayat, degraded land through SHGs	Seeds will be purchased by Uttaranchal Forest Development corporation	Provide additional employment and income to SHGs	

Table 15 Changing approaches to forest conservation

Approaches	Main features
Sustainable timber harvest	Timber harvest in amount equal to increment, thus capital remaining intact; replete with failures suggesting that ecosystems do not behave like banks
Harvest of non-timber forest products (NTFPs) and development of economic enterprises based on them	Expected to be more sustainable because trees are cut, only their parts such as resin, leaves, fruits, bark and organisms like mushrooms, lichens, and medicinal herbs collected. However, sustainable harvest practices are poorly developed and hardly practiced. Market is also poorly understood
Carbon sequestration in response to the global warming, an easily measurable ecosystem service	A part of Clean Development Mechanism of Kyoto protocol; but trade is still in infancy, at least in developing countries. Easy to trade because carbon is not required to be transported and quality of good invariant
Other ecosystem services, e.g., soil formation, pollination of crops hydrologic regulation, and regulation of biogeochemical cycle	Science of ecosystem services still in infancy; payment mechanism scarcely in place

Source Singh, S. P. (2007) Selling of ecosystem services. Samaj Vigyan Shodh Patrika. Special Issue (Uttarakhand-1). pp. 3–10

Table 16 A summary of ecosystem attributes and services of major forest types in central Himalaya and of *Lantana camara* that follows after deforestation

Forest	Attributes	Supply of ecosystem services
Banj Oak	• Large biomass (about 400–500 t ha^{-1}) • Fine roots and carbon deposition up to 1.50 m or more • High investment of photosynthate in ectomycorrhizae, massive yearly nutrient return to soil through litter fall	• Deep soil formation • High soil fertility • Effective carbon sequestration (also in deep soil) • Considerable nutrient and water retention
Chir pine (*Pinus roxburghil*)	• Small biomass (about 200 t ha^{-1}) • High productivity (12–20 t ha^{-1} yr^{-1}) • High nutrient use efficiency, great ability to tolerate stress • Effective colonization of mountain scopes and control over soil nutrients	• Supply of ecosystem services from steep slopes with little soil • Retention of nutrient on step and rocky slopes, associative nitrogen fixation (at least in some pines)
Alder (*Alnus nepalensis*)	• Small biomass (about 100 t ha^{-1}) • Very high productivity (20–30 t ha^{-1} yr^{-1}) • High effective colonization of landslips: High rate of N-fixation (up to 200 kg ha^{-1} yr^{-1})	• Development of tree cover (a sort of raining plantation by nature) without any cost • Facilitation of other species • Nutrient supply to other ecosystems.
Lantana Bush land (*Lantana camara*)	• Low biodiversity • Flammable by cool fire • Productivity similar to forests but low biomass, very shallow rooted	• Low soil C storage • Hydrologic cycle adversely affected • Persistent but light fire regime • Loss of biodiversity

Source Singh, P. (2007). Selling of Ecosystem Services. *Samaj Vigyan Shodh Patrika*, pp. 3–12

Table 17 Altitudinal variations in natural vegetation

Type	Height	Species
Sub-tropical	<1,400	Aonla, Semal, Jamun, Reetha, Mango, Malu, Tun, Buransh, Anyaar,
Sub-temperate	1,400–2,000	Pine, Oak, Buransh, Anyaar, Kafal, Payan, Bhimal
Temperate	2,000–3,300	Oak, Mawa, Pangar, Korak, Silver, Fur, Uttis, Thuner, Devrigaal, Ransal, Devdar, Thamrigal, Jumura
Alpine	>3,300	Grassland, Alpine meadows, Terang, Deodar, Bhojpatra, herbs

Source Compiled by the author

Glossary

Agriculture the occupation, business, or science of cultivating the land, producing crops, and raising livestock

Agronomy the science of soil management, land cultivation, and crop production

Bugyal is a nomenclature for alpine meadows

Cash crops a crop grown for direct sale rather than personal consumption

Danda highland forest area

Diversification the provision or development of greater variety

Ecology the study of the relationships and interactions between living organisms and their natural or developed environment

Emigration the act of leaving a native country to live in another country

Enhancement to improve or add to the strength, worth, beauty, or other desirable quality of something

Fragile easy to break, damage, or harm, usually because delicate or brittle

Gad or Gadhera is synonym to stream

Gangar lowland residential area

Immigration the act of people entering into a new country to settle permanently

Khet and Bari synonym to agricultural fields

Livelihood work done to earn a living or whatever provides a source of income

Niche any recess or hollow such as in a rock formation

Off-season a time of year when activity or business is at a low level

Sectoral a component of an integrated system such as an economy or a society

Van Panchayat elected members of the village assembly who look after the conservation of forest

V. P. Sati, *Towards Sustainable Livelihoods and Ecosystems in Mountain Regions*, Environmental Science and Engineering, DOI: 10.1007/978-3-319-03533-8,

About the Author

Vishwambhar Prasad Sati (b. 1966), gained his D.Litt. in 2011 and Ph. D. in 1992, in Geography, is a Professor of Geography and Resource Management, Mizoram University, Aizawl. He has above 20 years of teaching and research experience and he has served many national and international educational institutes in various capacities—Assistant Professor, Associate Professor and Professor in the Madhya Pradesh Higher Education and Associate Professor in the Eritrea Institute of Technology, N.E. Africa. He has been Visiting Scientist of the Indian National Science Academy (2012), Visiting Scholar of the Academy of Sciences for Developing Countries and Chinese Academy of Sciences (2010), worked at the IMHE, Chengdu, China; Fellow of the Indian Council of Social Science Research (2008–2009), worked at the HNBGU, Srinagar Garhwal; an Associate at the Indian Institute of Advanced Studies (2008), and a fellow at the G. B. Pant Institute of Himalayan Environment and Development (1993). He has completed six Research Projects, written 18 Reference and Text Books, published 74 Research Papers in journals of international and national repute and above 50 articles in magazines and newspapers, presented research papers in 17 countries, received fellowships and travel grants from 18 research organizations for participating in the various International Events, supervised five Ph.D. thesis, chaired academic sessions, participated as Resource Person in the national and international conferences, and serving many International professional bodies as member, editor and reviewer.

V. P. Sati, *Towards Sustainable Livelihoods and Ecosystems in Mountain Regions*, Environmental Science and Engineering, DOI: 10.1007/978-3-319-03533-8,

Printed by Printforce, the Netherlands